Zukunftstechnologien für den multifunktionalen Leichtbau

Series Editor
Open Hybrid LabFactory e.V., Wolfsburg, Niedersachsen, Germany

Ziel der Buchreihe ist es, zentrale Zukunftsthemen und aktuelle Arbeiten aus dem Umfeld des Forschungscampus Open Hybrid LabFactory einer breiten Öffentlichkeit zugänglich zu machen. Es werden neue Denkansätze und Ergebnisse aus der Forschung zu Methoden und Technologien zur Auslegung und großserienfähigen Fertigung hybrider und multifunktionaler Strukturen vorgestellt. Insbesondere gehören neue Produktions- und Simulationsverfahren, aber auch Aspekte der Bauteilfunktionalisierung und Betrachtungen des integrierten Life-Cycle-Engineerings zu den Forschungsschwerpunkten des Forschungscampus und zum inhaltlichen Fokus dieser Buchreihe.

Die Buchreihe umfasst Publikationen aus den Bereichen des Engineerings, der Auslegung, Produktion und Prüfung materialhybrider Strukturen. Die Skalierbarkeit und zukünftige industrielle Großserienfähigkeit der Technologien und Methoden stehen im Vordergrund der Beiträge und sichern langfristige Fortschritte in der Fahrzeugentwicklung. Ebenfalls werden Ergebnisse und Berichte von Forschungsprojekten im Rahmen des durch das Bundesministerium für Bildung und Forschung geförderten Forschungscampus veröffentlicht und Proceedings von Fachtagungen und Konferenzen im Kontext der Open Hybrid LabFactory publiziert.

Die Bände dieser Reihe richten sich an Wissenschaftler aus der Material-, Produktions- und Mobilitätsforschung. Sie spricht Fachexperten der Branchen Technik, Anlangen- und Maschinenbau, Automobil & Fahrzeugbau sowie Werkstoffe & Werkstoffverarbeitung an. Der Leser profitiert von einem konsolidierten Angebot wissenschaftlicher Beiträge zur aktuellen Forschung zu hybriden und multifunktionalen Strukturen.

This book series presents key future topics and current work from the Open Hybrid LabFactory research campus funded by the Federal Ministry of Education and Research (BMBF) to a broad public. Discussing recent approaches and research findings based on methods and technologies for the design and large-scale production of hybrid and multifunctional structures, it highlights new production and simulation processes, as well as aspects of component functionalization and integrated life-cycle engineering.

The book series comprises publications from the fields of engineering, design, production and testing of material hybrid structures. The contributions focus on the scalability and future industrial mass production capability of the technologies and methods to ensure long-term advances in vehicle development. Furthermore, the series publishes reports on and the findings of research projects within the research campus, scientific papers as well as the proceedings of conferences in the context of the Open Hybrid LabFactory.

Intended for scientists and experts from the fields of materials, production and mobility research; technology, plant and mechanical engineering; automotive & vehicle construction; and materials & materials processing, the series showcases current research on hybrid and multifunctional structures.

Klaus Dröder · Thomas Vietor
Editors

Future Automotive Production Conference 2022

Springer Vieweg

Editors
Klaus Dröder
Technische Universität
Braunschweig, Niedersachsen, Germany

Thomas Vietor
Institut für Konstruktionstechnik (IK)
TU Braunschweig
Braunschweig, Niedersachsen, Germany

ISSN 2524-4787 ISSN 2524-4795 (electronic)
Zukunftstechnologien für den multifunktionalen Leichtbau
ISBN 978-3-658-39927-6 ISBN 978-3-658-39928-3 (eBook)
https://doi.org/10.1007/978-3-658-39928-3

Die Deutsche Nationalbibliothek verzeichnet diese Publikation in der Deutschen Nationalbibliografie; detaillierte bibliografische Daten sind im Internet über http://dnb.d-nb.de abrufbar.

Springer Vieweg

This Springer Vieweg imprint is published by the registered company Springer Fachmedien Wiesbaden GmbH, part of Springer Nature.
The registered company address is: Abraham-Lincoln-Str. 46, 65189 Wiesbaden, Germany

Contents

Bio-based Material

Generative Manufacturing

Factories of the Future

Design and Simulation

Innovative and Smart Production

A Quantitative Method for the Investigation of Digitized Surfaces After Fine Milling Machining

Maik Mackiewicz(✉) and Jannik Backhaus

Volkswagen AG, Wolfsburg, Berliner Ring 2, 38440, Germany
{maik.mackiewicz1,jannik.backhaus1}@volkswagen.de

Abstract. Due to increasing efforts to shorten the time to market of new vehicle models, press tooling is under increasing competitive pressure. The manufacturing of press tools for high-quality body components depends on a large extent on the quality of the tools active surfaces as free-form surfaces. The use of non-contact optical measurement methods to digitize the tool surfaces creates high-density point clouds that have great potential to improve the manufacturing process through their analysis. In this paper, a method for the analysis of these point clouds is presented, which combines the qualitative and quantitative analysis of active surfaces, to enable conclusions about the process strategy and allow further optimizations to achieve cost and time savings. It is demonstrated on the example of a test specimen with characteristic elements of a press tool for vehicle body components. A parameter SQ is defined, which describes the area percentage inside of tolerance limits, acting as a quality parameter for the manufacturing result. The used test specimen achieves a SQ value of 93.9%. The robustness of SQ is investigated by a variation of the tolerance limits and a possible applicability to other components outside of vehicle body components is considered. Furthermore the distribution of the shape deviation is evaluated to determine whether too much or too little material removal causes the deviations, as well as in which angles it mainly occurs. It can be shown that the qualitative analysis of the false color image can be confirmed by the quantitative results of the proposed method. Based on this, optimization measures for the machining process are derived.

Keywords: 3D-metrology · Data analysis · Press tool manufacturing · Milling precision

1 Introduction

For the production of functional, high-quality vehicle bodies, high-quality body components are required, which must meet the highest standards of dimensional accuracy and surface appearance. The bodies of modern vehicles consist of more than 200

K. Dröder and T. Vietor (Eds.): *Future Automotive Production Conference 2022*, Zukunftstechnologien für den multifunktionalen Leichtbau, pp. 3–15, 2023.
https://doi.org/10.1007/978-3-658-39928-3_1

different components, most of which are produced by forming processes [1]. Press tools are required as press shop operating equipment for the forming of body components. The manufacturing of a tool set, consisting of up to six individual tools, can take up to 20 months. This has resulted in a need to shorten the manufacturing process for press tools, due to increasing efforts to shorten the time to market of new vehicle models [1, 2]. In general, new competitors in the global vehicle manufacturing industry and the relocation of large parts of the value chain to other countries or economic zones have created a need for process optimization throughout the entire manufacturing industry. The digitization of manufacturing processes represents an important approach to making processes more flexible and efficient, as well as ensuring their transparency through detailed information. This creates high potential for reducing the time and costs of manufacturing processes while at the same time increasing product quality [3]. To achieve these goals, new technologies are increasingly being used that are either data-based or generate large amounts of data and thus provide information about the manufacturing process. As the importance of product and process data for the optimization of the value chain will increase in the future and targeted analyses of such data allow valuable statements to be made about the production system, its individual elements and the process parameters, a method is needed to extract the complex, difficult-to-interpret information from the digital tool surface and make it usable [1, 4]. This paper therefore presents a method for generating additional information from the high-density point cloud of the digitized tool surface. It is examined how the manufacturing result can be made measurable and comparable in order to evaluate the current process. Furthermore, it will be discussed how qualitative observations from false color images can be verified and the resulting information can be transferred back into the machining process. The aim is to reduce the form deviation in future manufacturing processes of press tools and to prevent additional time and cost requirements in subsequent processes.

2 Background

The physical manufacturing of large press tools for vehicle body components consists of the casting of the structures with subsequent mechanical machining, heat treatment and a manual surface finish before the tool qualification for series production takes place in tryout. The materials used in this process depend on the planned total number of parts over the tools operating time, the required material and strength properties and the process-related complexity of the component to be produced. In general, steel castings such as EN-JS 1070 or EN-JS 2070 and rolled steels such as 1.2379 or 1.2320 are used. Mechanical machining is made up of various cutting processes, with roughing, pre-finishing and fine finishing being relevant for machining the tools active surfaces. Complex free-form surfaces are machined by ball or torus cutters traversing the surface in individual milling lines [1]. Fine finishing is performed with high-speed cutting (HSC) which, in addition to dry machining and very low thermal stress on the surface due to high heat transport through the chip, achieves a significantly larger metal removal rate than conventional milling due to high feed rates and speeds. As a result, HSC machining is ideally suited for fine finishing of the tool active surfaces due to high manufacturing precision and good economic efficiency and is state of the art in tool and die making [5]. The resulting surface texture and shape deviation are affected by the milling strategy, the cutting tool, used process parameters and the workpiece material, among other factors [1].

Shape deviation as profile cut	Type of deviation	Cause of origin
1st order: shape deviations 	straightness, flatness, roundness deviations, etc.	errors in machine tool control, deflection of the machine or the workpiece, incorrect clamping of the workpiece, hardness distortion, wear
2nd order: waviness 	waves	off-center clamping, form or running deviations of a milling cutter, vibrations of the machine or tool
3rd order: roughness 	grooves	shape of the tool cutting edge, feed or infeed of the tool
4th order: roughness 	scores, scales, domes	process of chip formation, material deformation during abrasive blasting, bud formation during galvanic treatment

Fig. 1. System of shape deviation orders [6]

According to DIN 4760, the totality of all deviations of the actual surface from the nominal geometric surface can be described by proportions of six orders of the shape deviation. In general, the first to fourth order shape deviations superimpose to form the actual surface. Its resulting surface texture and shape deviations affect many of its functions, such as wear, friction, sealing, assembly and visual appearance [6–8]. Figure 1 shows profile sections of the deviation orders relevant for the shape deviation. With increasing order, the corresponding deviation types and their cause of origin become finer in their dimension. Since the surface texture and shape deviations have a significant influence on the function of the surface, their accurate characterization is not only an important way to increase manufacturing accuracy, but also to guarantee the performance of the component [8]. This characterization is defined by Whitehouse as surface metrology [9]. The texture is the result of the process characteristics and the shape deviations are the footprint of the machine and the machining tool, including vibrations, geometric error motions and thermal distortion. The combination of surface texture and shape deviation forms the surface topography of the workpiece [8].

Before an examination of the surface can be carried out, it must be measured, which is possible in two dimensions. 2D measurement generates profile sections of the surfaces, which individually do not provide any information about the properties of the entire surface. Contact measurement methods (CMM) are used for this purpose, in which probes move over the surface and record the profile at the specific location. CMMs have the advantage of stable and reliable results, good repeatability and a wide range of applications with high precision. However, the probe geometry limits the accuracy of geometry elements such as deep grooves and small holes. Contact with the surface can also cause damage and affect measurement results. Especially for 3D measurement through many profile sections, CMMs require a lot of time. Since many surfaces affect the dimension accuracy of the drawn part during the production of press tools, the use of CMMs as a quality inspection would be extremely time consuming and cost-intensive and may even further prolong the process [7]. Noncontact measurement methods (NMM) are used for a 3D measurement, with the majority of NMMs using combinations of traditional optical metrology and information processing techniques. Mostly active measurement systems like laser triangulation or raster methods are used, which in contrast to passive measuring systems have special instruments to generate the measurement light or sound. Active measurement systems have the advantage of a high measurement accuracy and a wide range of supporting systems to reduce interferences such as secondary reflections, specular reflections, volumetric scattering and color differences on the measurement results. The overall advantages of NMMs are that all data in the image plane are obtained at the same time, no surface damage occurs, almost all materials can be measured and the measurement speed is high [7, 8]. Some NMMs, such as structured light projection, from the field of high-definition metrology (HDM) are able to digitize the entire workpiece surface in high resolution and generate a high-density point cloud. For this purpose, structured light is directed onto the workpiece surface in form of an array of straight stripes or a continuous spectrum of light that can be interpreted as an ensemble of finely sliced stripes. The pattern, distorted by the surface geometry, is then evaluated by a camera system and the surface is digitally reproduced as a point cloud [10].

An important area of application for structured light projection is the digitization of free-form surfaces in the context of automotive manufacturing. Since free-form surfaces in body components can meet the requirements for appealing designs in combination with aerodynamic shapes and ergonomic properties, digitized tool surfaces support press tooling as a key part of the vehicle production process [11, 12]. Depending on the extent of deviation from the nominal geometry, the complexity of the component and the interaction of departments involved, the manufacturing time can be shortened or, in worst case, noticeably extended [1]. The characterization of surfaces by observing first to fourth order shape deviations already provide a broad understanding of the influence of surface topography on their function [7]. Sun, Gao, Zhao and Tang investigate an approach to in-process surface texture condition monitoring to replace the surfaces textures inspection by the use of images of machined and cleaned parts with a back propagation neural network. The model determines the resulting surface texture by a correlation between texture image features and process signal features. They show that on-line determination of third and fourth order shape deviations is possible and can efficiently replace conventional image analysis [13]. Wang, Xi and Du developed a method to evaluate the 3D surface form error of large complex surfaces by using HDM. They process digitized point clouds of the workpiece surfaces into height-coded gray images and characterize them by entropy and contrast. Using the example of an engine block face, they show that an analysis of the high density point cloud provides a much deeper understanding of surface texture and deviation. They suggest that the information can be used to improve the surface machining process by performing process diagnostics [14].

Yin, Du, Shao, Wang and Xi present another method to perform surface analysis based on HDM. They investigate the sealing of face-milled surfaces using the example of a cylinder head and its counterpart, the cylinder block. They reconstruct the surface from a high density point cloud and determine channels in the sealing surface between the two components. In doing so, they can determine the channel direction as an important feature for the sealing of the contact surface. Based on this, they define a contact area percentage as a threshold value, above which the contact is considered sealed [15]. From the investigations it is evident that the analysis of surfaces allows improved knowledge about the part function and that a feedback of the gained information into the machining process can additionally optimize it. Due to the increasing use of HDM and the generation of high density point clouds, it can be shown that the analysis of the digitized surfaces in addition to the conventional methods further improves the understanding of correlations. Often third and fourth order shape deviations are extracted from the point clouds and investigated, so there is a need for methods in the first and second order range as well. For the manufacturing process of press tools, high-density point clouds offer opportunities to supplement qualitative analyses of the tool working surfaces, such as false color images, with quantitative parameters, as well as their feedback into the machining process.

3 Methods

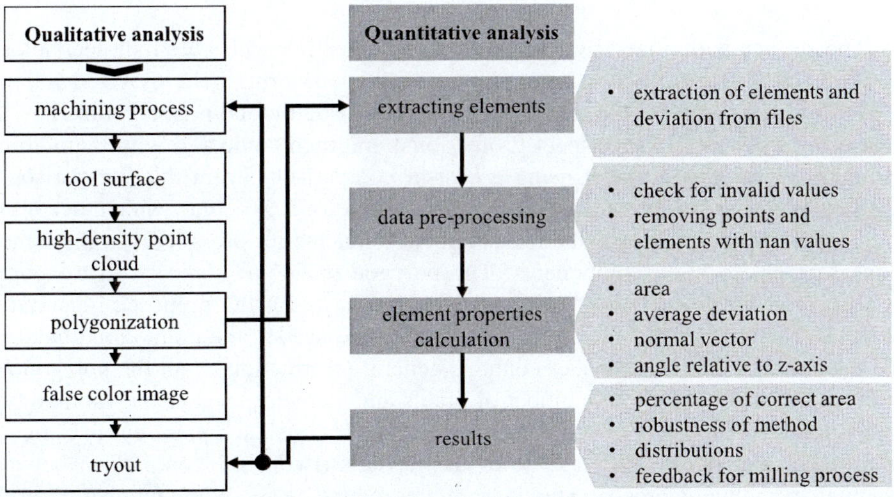

Fig. 2. Process chart of a qualitative and the proposed quantitative method

Figure 2 shows a process chart of a qualitative analysis that uses false color images to provide information for the tryout. Additionally, the proposed quantitative analysis based on the high density point cloud is shown. The proposed quantitative method is demonstrated on a test specimen with typical geometry elements of press tools for the automotive industry, such as different steep flanks that occur in door entries for side parts or concave and convex areas for controlling the material flow. The geometry of the test specimen is shown in Fig. 3. It has a length of 1240 mm, a width of 350 mm, a height of 250 mm and is made of EN-JS 2070 by sand casting, without further heat treatment. In order to exclude the influence of the surface layer properties on the machining result, 20 mm casting allowance is added on the entire surface. Machining consists of roughing to 1 mm, pre-finishing to 0,15 mm allowance and fine finishing in the HSC process. The entire HSC process takes place in three axis, so that the milling head is moved only in the x-, y- and z-direction. As a result, the cutting tool always maintains an orthogonal orientation to the surface planes and the same orientation as the z-axis.

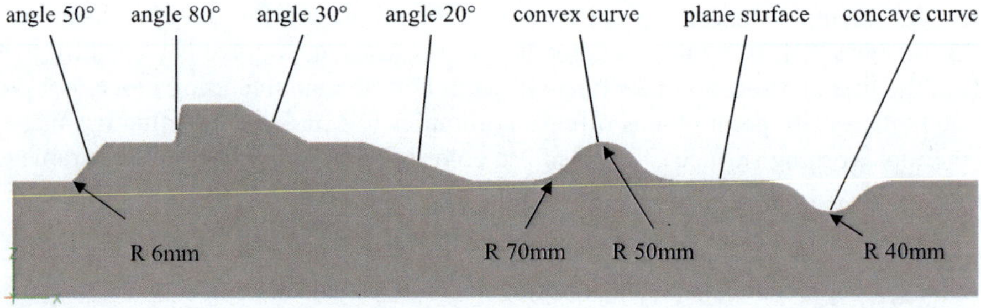

Fig. 3. Cross-section of test specimen used with typical geometrical elements

The surface is digitized by structured light projection with blue light technology. A GOM ATOS III Triple Scan hand scanner with a measuring field of 500 × 500 mm is used for this purpose. The resulting high-density point cloud is polygonized by the measuring software GOM Inspect Professional and then compared with the nominal geometry of the CAD model using a best-fit comparison. From this comparison, a false color image is generated with tolerance range of ±0.05 mm, which, according to the manufacturer, matches the measurement tolerance for press tools in the dimensions for vehicle body components. The polygonized point cloud is also exported in a polygon file format (.ply) and serves as the basis for the proposed quantitative analysis. The structure of the.ply file consists of the header, the point cloud data and the surface elements. The header contains general information about the polygonized point cloud, such as the total number of points and elements used or the variable type within the file. The point cloud data describes line by line each point by its Cartesian coordinates and the resulting value of the deviation from the comparison between nominal and actual geometry. The lines are interpreted as ascending numbering of the points, so that each point is assigned an ID. This ID is used by the surface elements to construct polygons. In this study, triangles are used to digitally recreate the surface.

In each row of the surface element part in the file, the number n of points needed for the specific elements is specified, followed by the n point IDs that make up the element. From this file structure, the surface elements can be extracted and further processed.

$$\vec{u} = \begin{pmatrix} x_{P2} \\ y_{P2} \\ z_{P2} \end{pmatrix} - \begin{pmatrix} x_{P1} \\ y_{P1} \\ z_{P1} \end{pmatrix} \tag{1}$$

$$\vec{v} = \begin{pmatrix} x_{P3} \\ y_{P3} \\ z_{P3} \end{pmatrix} - \begin{pmatrix} x_{P1} \\ y_{P1} \\ z_{P1} \end{pmatrix} \tag{2}$$

$$\vec{n} = \vec{u} \times \vec{v} \tag{3}$$

$$A_{triangle} = \frac{1}{2}\left|\vec{n}\right| \tag{4}$$

$$s_{avg} = \frac{1}{3}(s_{P1} + s_{P2} + s_{P3}) \tag{5}$$

$$\varphi = \cos^{-1}\left(\frac{\vec{n} \cdot \vec{z}}{\left|\vec{n}\right| \cdot \left|\vec{z}\right|}\right) \text{with } \vec{z} = \begin{pmatrix} 0 \\ 0 \\ 1 \end{pmatrix} \tag{6}$$

As shown in Fig. 2, the data is preprocessed by checking for invalid values. The deviation information can contain nan values, which arise if the digitized surface cannot find a corresponding counter surface in the surface comparison. This occurs, for example, in the case of holes, since the active surface is only a surface model in which the 3D geometry of the hole is not defined, while the CAD model is a solid model with defined hole geometry. Points at which a nan value of deviation occurs and elements containing these points are removed from the data. After that, properties of each element are calculated from the cleaned data by using two vectors $\vec{u}$ (Eq. 1) and $\vec{v}$ (Eq. 2) and their normal vector $\vec{n}$ (Eq. 3). In addition to the area of an element $A_{triangle}$ (Eq. 4), the deviations s_P of its three points, which describe the distance of the measured points from the nominal position in the direction of $\vec{n}$, are averaged (Eq. 5). Furthermore the angle φ of the element relative to the z-axis is calculated (Eq. 6), since this is the position of the milling cutter during milling [16]. Thus, the orientation of the surface relative to the cutter during machining is determined by this. The elements properties are used to calculate the percentage of the surface within the specified tolerance of the shape deviation. This parameter is named surface quotient (SQ) which relates the area within the tolerance A to the total area of the surface A_{total}.

$$SQ = \frac{A}{A_{total}} \cdot 100\% \tag{7}$$

SQ serves as a quantitative parameter for measuring the manufacturing result, making the results of different manufacturing processes comparable. Since SQ depends on the chosen tolerance range, its variation is examined and interpreted as robustness of the method. Due to the determined element properties, investigations of their distributions as characteristics of the manufacturing result become possible. For this purpose, the distribution of areas outside the tolerance in ranges of the angle φ between the surface and the cutter is exemplarily investigated. Based on the gained information, a proposal for the optimization of the cutting process is given.

4 Results

Fig. 4. False color image after surface comparison between digitized surface and CAD model

The machining result of the test specimen is shown as a false color image in Fig. 4. It can be noticed that the deviation in the partial areas of the flanks with 50° and 80° as well as in the concave curve lie outside of the set tolerance of ± 0.05 mm. In comparison, no deviation from the tolerance can be determined in the area of less steep flanks, the convex curve and the planes.

Table. 1. Basic information about the digitized surface

Information	Value
Number of points in point cloud	311.155
Number of elements	619.369
Number of nan values	0
Number of elements containing nan value points	0
Total area	479.350 mm^2

Before the calculation of element properties is done, the basic information of the digitized surface is shown in Tab. 1. During pre-processing, zero points with a nan value are detected. Therefore no points or elements have to be deleted.

Table. 2. Surface quotient and analysis of failed area

Information	Value (%)
Surface Quotient	93.9
Percentage of area outside of tolerance	6.1
Percentage of area below tolerance	26.7
Percentage of area above tolerance	73.3

The calculation of the surface quotient shows that 93.9% of the surface falls within the tolerance range. The remaining 6.1% are above or below the set tolerance limits. A more detailed evaluation reveals that 26.7% are below tolerance and thus too much material has been removed, while 73.3% are above tolerance and not enough material has been removed. The results are presented in Tab. 2. Since SQ depends on how the two tolerance limits are set, they are varied in a range of 0 to 0.30 mm and −0.30 to 0 mm respectively to investigate the change in SQ.

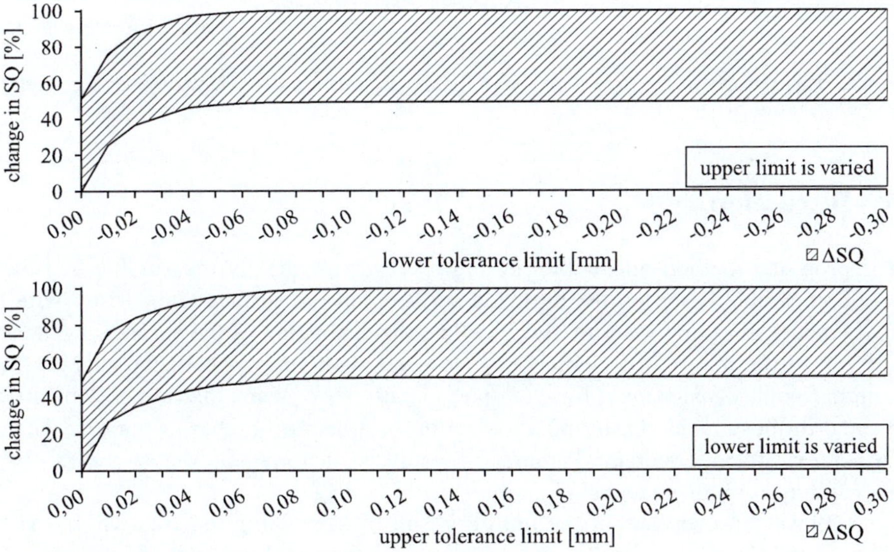

Fig. 5. Range of SQ due to variation of one tolerance limit

Both tolerance limits are varied by keeping one constant while the other varies. The result of this investigation can be seen in Fig. 5. The change in SQ is shown in each case as a band, the width of which shows the intensity of the change due to a limit variation. It can be observed that the width of the band with a variation of the upper limit is similar to that of the band with a variation of the lower limit. As explained at the beginning, it can be observed that deviations of the surface from the nominal geometry occur in areas that have a steep angle relative to the planes, such as the flanks with 50° and 80° or the concave curve. In order to verify this qualitative observation with the proposed method, the distribution of the element areas with an average deviation outside the tolerance over the angles to the z-axis is examined. Figure 6 shows the corresponding distribution of the area percentages. The result reveals that with increasing angles of the surface normal to the z-axis, the proportion of the total surface outside the tolerance limits increases. This trend can be observed consistently except for the range of 50 to 60 degrees, until it reaches its maximum of over 90% at 80 to 90 degrees. The range from 40 to 50 degrees has a higher percentage, due to the deviations in the concave curve.

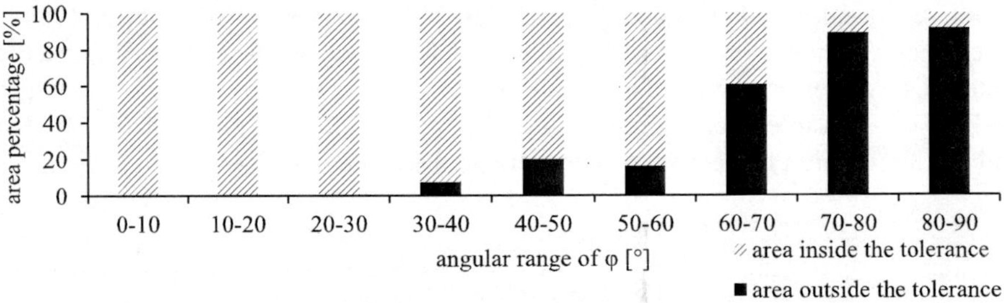

Fig. 6. Distribution of the area percentage inside and outside the tolerance over the angles to the z-axis

5 Discussion

The proposed method makes the milling result quantitatively measurable and therefore comparable based on the digitized surface and the surface comparison with the CAD model. It is important to always relate the evaluation of the surfaces to the area and not to the number of elements, since areas with strong geometry changes have a higher point density than, for example, planes. This means that more elements are formed in these areas during polygonization in order to be able to map the geometry with sufficient accuracy. Parameters would be distorted by this. An area related approach is normalized and comparable.

SQ gives an overview of the milling result by describing the area within the tolerance limits. An investigation of the area outside the tolerance limits allows a further characterization of the milling result by determining the distribution of the area above and below the tolerance. Areas where too little material has been removed cause increased grinding efforts, while areas where too much has been removed have to be filled by additive processes. Both cases cause additional time and cost efforts in the subsequent processes for tool qualification for series production. The reason why too much or too little material has been removed are manifold and depend on the process used. For example, non-optimal milling parameters, an unsuitable cutting angle or an unfavorable load on the cutting tool can be causes for deviations. The determination of the distribution of surfaces above and below the tolerance limits, as shown in Tab. 2, allows an initial feedback of information into the machining process and the definition of suitable process adjustments to improve it, for example a change in the milling strategy.

The robustness of SQ is represented by a limit variation in Fig. 5. When the method is applied to the surface of the test specimen, the change in SQ is similar when both tolerance limits are varied. The variation in the mentioned range causes a change in SQ up to 50%, depending on the combination of tolerance limits. Therefore, the choice of tolerance limits is crucial for a reliable SQ. The manufacturing of press tools, on the other hand, is about transferring the surface of the CAD model into the physical tool as precisely as possible, so that the perfect result would be a deviation of 0. For this reason, symmetrical tolerance limits around a deviation of 0 are suitable for this specific application, whereby the smallest possible distance between the limits is determined by the measuring accuracy of the optical measuring system. However, SQ can also be applied to one-sided tolerances or different tolerances in positive or negative direction by using asymmetric tolerance limits. Exemplary, toleranced components where the shape deviation in a certain direction would affect the components function, while a deviation in the other direction would have less or no influence, be mentioned.

In addition to the information on the areas and their distribution above and below the tolerance limits, the proposed method also allows investigations between properties of the elements. To quantitatively verify the qualitative observation from the false color image in Fig. 4 that deviation tends to occur in areas with a large angle to the z-axis, it is useful to consider the percentage of area outside the tolerance as a function of angle, as shown in Fig. 6. The area percentages are calculated in 10 degree increments for angle ranges of φ from 0 to 90 degrees. It can be seen that as the angle φ increases, the proportion of the area outside the tolerance also increases, except for the range of 50 to 60 degrees. A possible explanation is that the deviation occurring in curved areas has a different formation pattern than in uniform flanks, since the concave area has deviations in the range of 40 to 50 degrees. Thus, the qualitative observation from the false color image can be confirmed with the quantitative result of the proposed method. This confirmed information can now be transferred back into the machining process. To reduce the deviation in areas with larger angle φ, the machining process can now be specifically checked and adjusted in the critical areas. Due to the three-axis machining of the test specimen and the resulting fixed position of the cutting tool, a logical approach in areas of large angles φ is to change the angle of the cutting tool in order to achieve either a better cutting angle or a more advantageous load on the cutting tool. The latter can be realized, for example, by a larger axial component and a lower lateral component of the load due to a different angle of the cutting tool in combination with a greater rigidity at an axial load of the tool holder.

6 Conclusion and Future Work

In this paper, a method for the analysis of digitized surfaces after fine milling machining, consisting of polygonized high-density point clouds has been presented. By calculating the properties of elements, new investigations are possible, which provide information to optimize the machining process. After a pre-processing of the point cloud, where points with a nan value as deviation and elements using these points are removed, specific properties of the elements are calculated. In addition to the area and the average deviation s_{avg}, the angle φ of the element to the z-axis is also determined, which is similar to the cutting tool position during machining.

SQ describes the percentage of the area within the set tolerance limits. A deeper analysis of the distribution of the area outside the tolerance limits offers a first opportunity to select specific optimization measures according to the focus of the deviation. These can be based on whether too much or too little material is removed.

A variation of the tolerance limits shows that the chosen limits are decisive for a reliable SQ. However, this also allows a wide use of the proposed method, since both symmetrical and asymmetrical tolerance limits can be chosen, depending on the influence of the shape deviation on a component function. In addition, an examination of the element properties in relation to each other offers the possibility of obtaining further information about the characteristics of the shape deviations. The distribution of the area outside the tolerance limits over the angular ranges of φ confirms the qualitative observation from the false color image that more deviations occur for surfaces with a larger angle φ.

Overall, the proposed method offers the advantage of making the manufacturing result measurable and comparable on the one hand and characterizing the shape deviation in such a way that efficient optimization of the machining process becomes possible on the other.

The increasing use of high-density point clouds in manufacturing metrology creates the need to accurately characterize further manufacturing results. It is also necessary to investigate more complex algorithms for the calculation of SQ. For example, additional limits can be used and the resulting areas can also be weighted differently. This further increases the applicability of the proposed method. With measurement methods that are able to additionally capture the third and fourth order shape deviations, analyses that also include the surface appearance become possible. This results in the possibility of investigating an improved machining process that produces smaller shape deviations and at the same time better surface appearances, which has a positive effect on subsequent processes in the manufacturing of press tools. Furthermore, suitable methods and structures for information feedback into the machining process have to be identified. Intelligent analysis methods such as machine learning models could be applied here.

Disclaimer
The results, opinions and conclusions expressed in this publication are not necessarily those of Volkswagen Aktiengesellschaft.

References

1. Birkert, A.R., Haage, S., Straub, M.: Umformtechnische Herstellung komplexer Karosserieteile. Springer, Berlin(2013)
2. Kluge, S.: Prozesse der Blechumformung. Hanser, München (2020)
3. Damasky, J., Cameron, R., Miltenburg, P.: Zielbild Automobilproduktion VDA 6000. Verband der Automobilindustrie e. V., Berlin (2021)
4. Damasky, J., Cameron, R., Miltenburg, P.: Smart Data in der Automobilproduktion VDA 6100. Verband der Automobilindustrie e. V., Berlin (2020)
5. Denkena, B., Tönshoff, H.K.: Spanen. Springer, Berlin (2011)
6. DIN 4760:1982-06: Gestaltabweichungen – Begriffe, Ordnungssystem
7. Du, S., Xi, L.: High definition metrology based surface quality control and applications, Springer, Singapore (2019)
8. Gao, W., Haitjema, H., Fang, F.Z., Leach, R.K., Cheung, C., Savio, E., Linares, J.: In-machine and in-process surface metrology for precision manufacturing. CIRP Ann. **68**(2), 843–866 (2019)
9. Jiang, X.J., Whitehouse, D.J.: Technological shifts in surface metrology. CIRP Ann. **61**(2), 815–836 (2012)
10. Iizuka, K.: Engineering optics, Springer, Cham (2019)
11. Savio, E., d. Chiffre, L., Schmitt, R.: Metrology of freeform shaped parts, CIRP Annals. **56**(2), 810–835 (2007)
12. Liu, S., Zhang, M., Kadam, P., Kuo, C.: 3D Point cloud analysis, Springer, Cham (2021)
13. Sun, H., Gao, D., Zhao, Z., Tang, X.: An approach to in-process surface texture condition monitoring. Robot Comput. Integr. Manuf. **48**(1), 254–262 (2017)
14. Wang, M., Xi, L., Du, S.: 3D surface form error evaluation using high definition metrology. Precis. Eng. **38**(1), 230–236 (2014)
15. Yin, Y., Du, S., Shao, Y., Wang, K., Xi, L.: Sealing analysis of face-milled surfaces based on high definition metrology. Precis. Eng. **73**(1), 23–29 (2022)
16. Papula, L.: Mathematik für Ingenieure und Naturwissenschaftler. Springer, Wiesbaden (2018)
17. Biondani, F., Bissacco, G., Hansen, H.N.: Surface topography analysis of ball end milled tool steel surfaces. Procedia CIRP **87**(2), 153–158 (2020)
18. Claus, F., Hamann, B., Leitte, H., Hagen, H.: Decomposing deviations of scanned surfaces of sheet metal assemblies. J. Manuf. Syst. **61**, 125–138 (2021)
19. d. Chiffre, L., Lonardo, P., Trumpold, H., Lucca, D.A., Goch, G., Brown, C.A., Raja, J., Hansen, H.N.: Quantitative Characterisation of Surface Texture. CIRP Ann. **49**(2), 635–652 (2000)
20. Fang, Z., Sugita, N.: Towards understanding and controlling of the surface texture patern in 5-axis ball-end milling using fast texture simulation. Precis. Eng. **74**, 80–91 (2022)
21. Oesterling, P., Heine, C., Janicke, H., Scheuermann, G., Heyer, G.: Visualization of High-Dimensional Point Clouds Using Their Density Distribution's Topology. IEEE Trans. Visual Comput. Graphics **17**(11), 1547–1559 (2011)
22. Hehenberger, P.: Computerunterstützte Produktion. Springer, Berlin (2020)

Technologies for Circularity

Fatigue Life of Refurbished Fiber Reinforced Thermoplastics

Dennis Weintraut[1](✉), Florian Kraft[1], Justus Freeden[1], and Robert Meltke[2]

[1] Fraunhofer Institute of Machine Tools and Forming Technology – IWU, Hermann-Münch-Street 2, 38440, Wolfsburg, Germany
{dennis.weintraut,florian.kraft, justus.freeden}@iwu.fraunhofer.de

[2] Fraunhofer Institute of Machine Tools and Forming Technology – IWU, Reichenhainer Street 88, 09126, Chemnitz, Germany
robert.meltke@iwu.fraunhofer.de

Abstract. This study contributes basic knowledge in the field of fatigue testing for refurbished fiber reinforced thermoplastics (FRTP). In a first step quasi-static tensile test were performed. From the tensile tests stress amplitude levels are retrieved for fatigue tension-tension test with a stress ratio of $R = 0.1$. Fatigue tests for four load levels were performed and S-N-curve for the virgin material was derived. In the next step, again fatigue tension-tension tests were performed but the maximum load was reduced to 20% of static strength and the material tested up to 1.2E6 cycles, to simulate one life cycle. After the long life fatigue tests the specimen were refurbished. The process was defined by using a heated press. The specimen were placed in a tool to keep the shape and then temperature and pressure were applied on the specimen, to melt the plastic and close possible delamination or micro cracks in the matrix material. The refurbished specimens were again fatigue tested on the same stress levels which were applied to the virgin material. It could be shown that a refurbishment of long life fatigue material and an evaluation of an S-N-curve for refurbished material is possible. Especially for high cycle fatigue, the S-N-curves are in a close range so that the use of refurbished material seems possible.

Keywords: Circular economy · Continuous fiber reinforced thermoplastics · Refurbishment · Fatigue analysis

1 Introduction

Composites, as glass and carbon fiber-reinforced plastics (GFRP, CFRP), substitute metallic materials in a continuously growing number of industrial applications due to their excellent specific mechanical properties leading to a further increase of composite use in the next years [1, 2]. In the automotive industry so far, composites

K. Dröder and T. Vietor (Eds.): *Future Automotive Production Conference 2022*, Zukunftstechnologien für den multifunktionalen Leichtbau, pp. 19–29, 2023.
https://doi.org/10.1007/978-3-658-39928-3_2

are only been established to a limited extent, in particular due to the high material costs and complex component production. However, advances in research and development have reduced cycle times and costs, which has increased the attractiveness of fiber-reinforced plastics in the automotive industry. A key advantage here is the use of impregnated thermoplastic sheet semi-finished products, so-called organic sheets. By heating and pressing processes organic sheets can be formed, thus reducing cycle times as no curing reactions occur, unlike it is found in thermosetting materials.

Against the background of the increasing focus on the circular economy, end-of-life (EoL) strategies for composites are also becoming increasingly important. While in the past composite parts and materials ended on a landfill or were thermal recycled new ideas are required due to political and social pressure for more sustainable emission reduction. The production of virgin carbon fibers, for example, have a 14 times higher energy consumption than steel [3]. This energy is wasted if the CFRP material is disposed after EoL. If carbon fibers are recycled 80 to 90% less energy is consumed compared to newly produced fibers [4]. However, CFRP and GFRP are difficult to recycle due to the combination of polymer matrix and reinforcing fiber. Current recycling processes are based on the separation of fiber and matrix or on the shredding of the EoL composite material. But the result of these recycling processes often means a degradation of material properties for example by cutting, grinding, solvolysis or pyrolysis [5]. Additionally, pyrolysis and solvolysis are not used in the field of thermoplastic composites due to high process costs. Instead, the EoL components are grinded, compounded and used as short fiber reinforcement in injection mold parts. Even if the material is recycled, there is a strong reduction of properties due to fiber shortening resulting in a loss of value. Therefore, recycling of composites often means downcycling.

A different approach for sustainable and circular solutions for EoL composite parts is reuse. In the field of composites, a distinction can be made between component reuse and structural reuse. Component reuse is defined as the usage of complete parts or assembly after a product reaches its EoL in another product. The parts need to be tested by non-destructive testing and if approved they can return into service. If no return into service is possible, they have to be recycled. In contrast to conventional recycling, structural reuse does not involve separation of fiber and matrix, nor does it involve shredding, which preserves the structural integrity of the composite material. The service life of the material is extended by reusing structural elements such as beams and panels. This concept has already been explored for thermosetting composites and has a high potential. By using thermal expending particles beams, panels or layers can be recovered from parts and be reused again [6]. The structural reuse approach can also be applied for thermoplastic composites. However, unlike thermosetting materials, the properties of the thermoplastic matrix allow a refurbishment of EoL structures. The differences between component reuse, structural reuse/refurbishment and recycling are shown in Fig. 1.

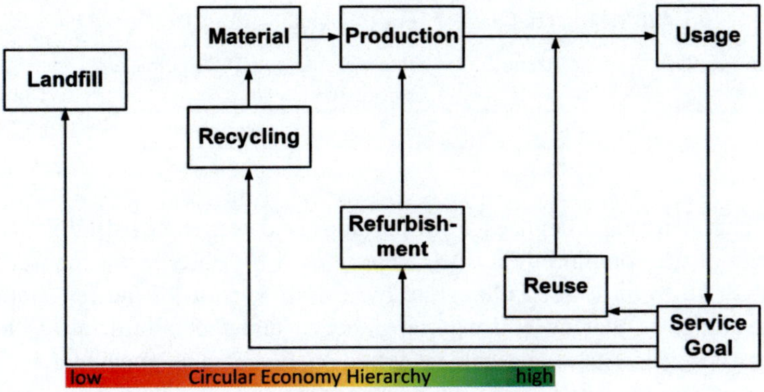

Fig. 1. Reuse vs. Refurbishment vs. Recycling and its circular economy hierarchy

By refurbishing organic sheets, it is possible to conserve the material properties of the composite and reduce the energy consumption. Although there are already isolated studies in FRTP reforming [7], the refurbishment approach has not been in focus yet. Several problems need to be discussed before refurbishing a part.

In general, parts are designed, simulated and tested for one life cycle. With structural reuse through refurbishment processes, the simulation and design will be more complex. Materials are already loaded for a number of cycles, which can lead to in-situ damages. Additionally composite materials are often joined with metals, for example, to form a hybrid component. To enable refurbishing of the organic sheets, these hybrid material compounds must be separated. Even if first studies of separating processes are available, for example Freeden et al. [6], it is not yet studied how the separating process affects the material. Before studying the effect on material properties by separating hybrid materials, it needs to be evaluated how the material properties of refurbished materials are, after reaching the design end-of-life. Goal of this study is to evaluate the fatigue life of refurbished material. To study refurbished organic sheets, specimens or components need to be tested under cyclic loading, refurbished and tested again to evaluate the fatigue properties of such materials.

2 Material and Methodology

In this study, a Tepex® Dynalite 102-RG600(x)/47% organic sheet with a thickness of 2 mm was used. The weave was a glass fiber twill 2/2 with a fiber volume content of 47%. As matrix material a polyamide 6 (PA6) was used. The material properties of the organic sheet are shown in Table 1.

The material was chosen due to its wide use in the automotive sector and the material properties of the thermoplastics. The matrix system is reversible meltable thus it can be refurbished by temperature and pressure.

Table 1. Material data of Tepex® Dynalite 102-RG600(x)/47%

Tensile modulus	Tensile strength	Fiber volume content	Ply thickness	Melting Temperature
18 GPa	380 MPa	47%	0.5 mm	220 °C

The specimen shape deviates from the approved test standard DIN EN ISO 527 [8]. Instead of the common three specimen types, a fourth type was chosen. It is recommended due to its lower failure rate by a draft version of the test standard. The major advantage is that no end caps are needed and the failure rate is lower than any other type [9]. The shape can be found in Fig. 2. The specimen is straight in the clamping and measuring area but curved outside of it. Quasi-static pretests have shown that all tested specimen led to valid results and no failure in the clamping area occurred.

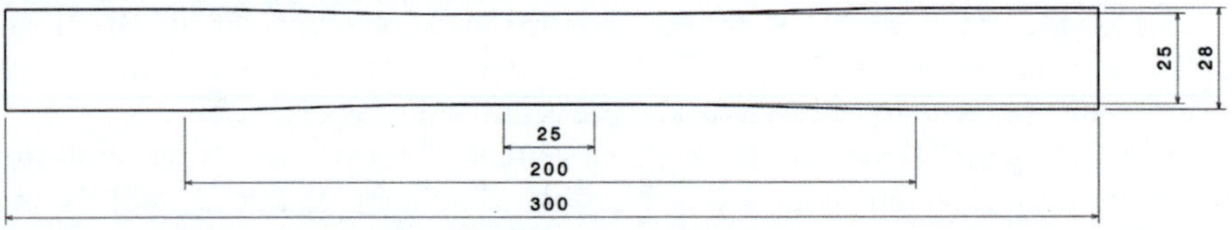

Fig. 2. Shape of specimen type 4 in accordance with [9]

Additionally, a small number of fatigue tests were preformed to validate the usage of the specimen shape. Different load levels were applied and the specimen tested until final failure. All tests led to valid results by failing in the measuring area and no failure outside the recommended area. A major advantage, concerning the tests of the refurbished material. Invalid tests need to be minimized due to the time consumption of the long life fatigue test and the refurbishment process. Furthermore, the specimen design has a major advantage if compared to the traditionally used specimen shapes concerning the chosen refurbishment process. For this process, it is necessary to apply a constant pressure and temperature above the specimen. Only then, a reconsolidation of the organic sheet can be provided.

To define the fatigue life of materials, cyclic loading tests need to be performed to evaluate an S-N-curve. In this study only a tension-tension test is performed in accordance with ISO 13003 with a stress ratio of $R = 0.1$. The tests were performed on a Zwick/Roell HA 100 dynamic testing machine. Four stress levels were defined: 35%, 50%, 60% and 70% of static Ultimate Tensile Strength (UTS_s). The testing frequency was set to 7 Hz; pre-tests have shown that the temperature increase is below the maximum of 10 °C, which is required by the test standard, while the test duration is still in an acceptable time. The four stress levels were tested up to final failure. Additionally, 20 specimens were loaded by 20% UTS_s up to 1.2 million cycles to simulate a life cycle of the material. Though in literature, at a value of 2.0 million cycles, long life fatigue strength is assumed, the value is decreased to reduce testing time and based on the performed fatigue tests prior to the long life fatigue tests. Additionally, it can be assumed that at this point the material is damaged but no fiber failures occurred which is necessary for a refurbishment. The long life fatigue tested specimens were refurbished in a heated press after their life cycle. For this purpose specimen were placed in a tool, shown in Fig. 3, to keep the shape. The press was heated up to melting temperature of the PA6 and closed after placing the tool and specimen. The temperature was increased to melting temperature and pressure was applied for 10 min and then cooled down for another 10 min below melting temperature. After removing manually the specimen by applying pressure, five specimens each defined stress level are again fatigue tested.

Fig. 3. Specimen in tool before refurbishment process

3 Experimental Evaluation

To define the loading levels quasi-static tensile test were performed. The values provided by the manufacturer could be confirmed and the load levels displayed in Table 2 were defined.

Table 2. Load levels

% of UTS	Stress in MPa
70	266
60	228
50	190
35	133
20	76

In total five specimens per stress level were tested and the S-N-curve was plotted. In Fig. 4, the S-N-curve for the virgin material is shown. It can be seen that the load levels are in a good range if compared to the numbers of cycles. The highest load level of 70% has a mean number of cycles to failure of 3289 cycles. This is a good range due to scattering of the material properties and that the testing machine requires some cycles to maintain the defined test parameters. The lowest load level of 35% has a mean value of 575,943 cycles to failure. To gain better results the lowest load level should have been reduced to 30% for getting closer to 1,000,000 cycles, which is common for fatigue testing. For long life fatigue testing a load level of 25% would be optimal, because it is close to 1.2 million cycles to failure. Due to the scattering of the numbers of cycles to failure in the field of high cycle fatigue, the chosen level of 20% is in an acceptable range. It needs to be ensured that no specimen fails before reaching the target of 1.2 million cycles and no fiber failure occurs based on the required time of long life cycle fatigue tests. The criteria of no occurrence of fiber failure is necessary because only matrix failures or delamination can be refurbished.

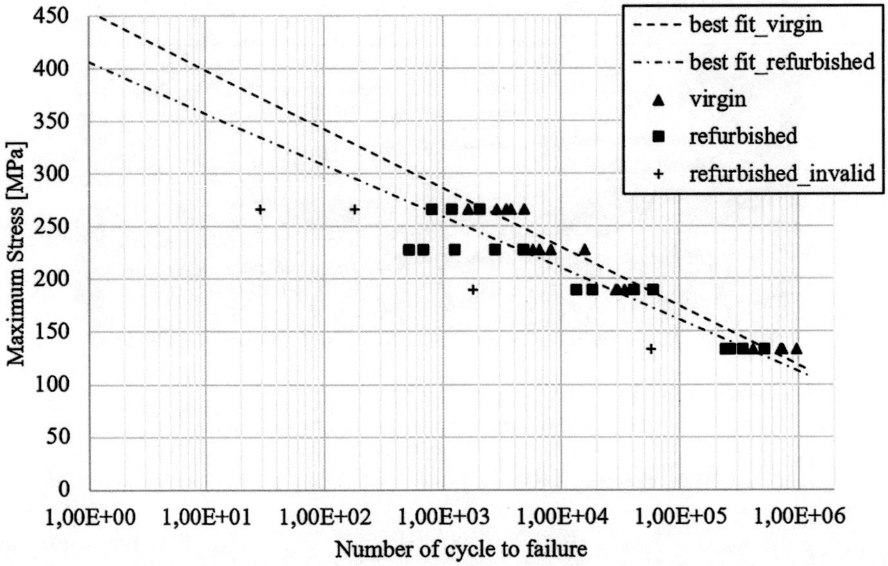

Fig. 4. S-N-curve of the virgin vs. refurbished material tested under T-T-loading

For the refurbishment process, a heated press was used to melt the thermoplastic matrix above melting temperature. Due to its low viscosity a tool were cut with the shape of the specimens, to prevent a change in the shape caused by loss of matrix material. The steel sheet has the same thickness as the specimen to guarantee a sufficient area pressure above the specimen. The press and the tool are heated up to 220 °C, the melting temperature according to the manufacturer. The tool is heated to prevent residual stresses caused by a sudden temperature change. After reaching the target temperature, the specimens are placed inside the tool and the press is closed. A pressure of 12.5 bar is applied for ten minutes on the samples. For another 10 min, the pressure is kept and the temperature reduced to 180 °C, that the thermoplastic is consolidated. Then the press is opened and the tool removed from the press. After another cooling period, the specimens were removed manually by applying a pressure on its surface. The duration of the second cooling period was not defined, it was necessary so that the specimens could manually be removed from the metallic tool. The refurbishment process parameters with the heated press is shown in Fig. 5.

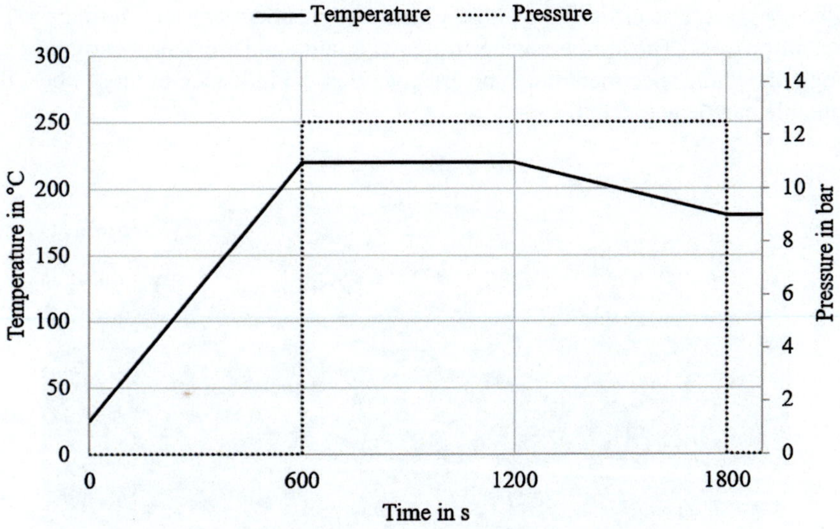

Fig. 5. Temperature and pressure vs. time for the refurbishment process

The specimens were measured and weight to compare the differences between the virgin and refurbished specimens, the results can be seen in Table 3. As is shows the refurbished specimen are losing mass by a simultaneous increase of its thickness. It is a minimal increase in thickness, which can be explained by deviations in the measuring position of the caliper gauge.

Table. 3. Masses and deviations of virgin and refurbished specimen

		virgin	Refurbished	Absolute deviation	Relative deviation (%)
Mass	Mean	29.04 g	28.34 g	0.7 g	2.47
	Maximum	29.50 g	28.89 g	1.16 g	4.10
	Minimum	28.48 g	27.65 g	0.36 g	1.28
	standard dev	0.38 g	0.32 g	0.25 g	0.90
Thickness	Mean	2.02 mm	2.05 mm	0.03 mm	1.51
	Maximum	2.06 mm	2.15 mm	0.15 mm	7.58
	Minimum	1.97 mm	1.96 mm	−0.08 mm	– 3.90
	standard dev	0.03 mm	0.06 mm	0.07 mm	3.36

The decrease of mass can be explained by the loss of matrix material during the refurbishment process. As it can be seen in Fig. 6 the matrix leaks during the refurbishment process, which was not prevented by the tool. The leak occurred on the upper side of the specimen only, which implies that the upper part of the press must be hotter than the lower part and the material is melting faster on the upper surface than on the bottom. This leads to undulation, as shown in Fig. 6, additionally. The bottom surface had no undulation. Probably only the top layer has undulation and all other layers are undamaged, this assumption could not be clarified and needs to be validated by a CT-scan. It can also be obtained that fibers are lost during the refurbishment process. Those fibers were in the edge area and were no continuous fibers through the whole specimen. It can be assumed that the influence of those fibers of the fatigue life can be neglected.

Fig. 6. Undulation and leaked matrix in refurbished specimen

After this, the four stress levels were tested under tension-tension fatigue loading and the S-N-curve evaluated as shown in Fig. 4. For the refurbished specimen only four results were invalid for the S-N-curve due to the deviation from the mean values. For 70% UTS two specimen are far below the minimum of 1000 cycles. For the load levels 60% and 35% UTS one specimen each is invalid. The rest of the samples are in a good range and can be seen as valid. Thus shows, that the chosen specimen shape can also be used for fatigue testing. Additionally the results are similar compared to the virgin specimen. The curves a nearly parallel, with a similar slope and axis intercept. Comparing the interpolated equations of both curves it can be obtained that after reaching around 9,400,000 cycles the two curves intercept. It could be assumed that the refurbished material can gain more strength after this point, but it is probably the result of the curve fitting and due to scattering in the results. An improvement caused by refurbishment seems more or less unlikely but should be a topic for further research.

4 Discussion

It could be shown that a refurbishment of loaded two-dimensional FRTP specimen is possible by using a heated press. A two-dimensional specimen shape was chosen to validate the refurbishment process; still a metallic tool was required to maintain the specimen shape during the process. Nevertheless, there was a loss of matrix material and undulation occurred on the upper surface probably caused during the closing process of the press. By using a continuous press, for example, a double-belt press, a consistent pressure and temperature can be applied on the upper and lower surface from the beginning to prevent undulation and material loss caused by the closing procedure and improve the speed in sights of an industrial serial refurbishment process. Furthermore, the refurbishment process needs to be qualified for more complex shapes and designs. It was used for two-dimensional specimen only, but not for three-dimensional structures. Although, it is necessary to evaluate if the shape can be kept for a those more complex designs and if the tools of the production process can be used or if another tool is required. Additionally no hybrid design was studied, for example an organic sheets combined with injection-molded thermoplastics. A refurbishment process needs to be evaluate for this material mix and it is required to study how the process affects the fatigue-life of the refurbished hybrid part. Nevertheless, it could be proven that a refurbishment of thermoplastic organic sheets is possible and two-dimensional geometries can be used again to reduce waste or downcylcing by shredding continuous fibers.

The study additionally showed that a refurbishment is not only possible for the chosen material; the material properties are still in a good range compared to the properties of virgin material. Especially for high cycle and very high cycle fatigue, the study showed that it could be used again. Furthermore the standard deviation of the refurbished material is smaller, for nearly each load level, only for 50% UTS the standard deviation is higher (20,782 to 3618 cycles). Thus indicates again that fatigue life testing can be used for refurbished materials. For low cycle fatigue on the other hand, further improvement of the testing is required. Especially the results of load level of 70% UTS show that invalid results can occur. Two of five tests were invalid, one nearly after the start. It could be a result of the undulation in combination with not proper set PID-control values that led to overshooting in the first cycles and causing damages to the specimen, even that the set values were kept constant between the virgin and refurbished material tests.

5 Conclusion

The carbon footprint of fiber-reinforced plastics is a major disadvantage for its use. While it can decrease the mass of a product and thereby its energy consumption during its life, the low recyclability compensates this effect negatively. In the present work, a new approach for a circular use of FTRP is presented. The refurbishment process takes advantage of the reversible melting capability of thermoplastics and provides a possibility to use FRTP again. It could be shown that the fatigue life of refurbished material is nearly as good as virgin material and by using a heated press a consolidation of the FRTP is possible. Nevertheless, it needs to be mentioned that the applied loads for the long life fatigue was below the loads for fiber failure. It needs to be validated how the behavior changes in case of those and if a refurbishment can be performed. In case of positive results, non-destructive testing methods need to be developed for this case. Only if refurbished materials without major inspections is provides a monetary benefit compared to virgin material a widely usage in the industry is possible. Not yet regarded in this study, due to the time consuming tests, was how big the influence of the refurbishment process is on the material. Therefore, further tests are required to load specimen up to 1.2 million cycles and test them under the four defined load levels but without a refurbishment. All three curves should be compared to gain a better understanding.

The last point, which needs to be verified, is the energy consumption of the refurbishment process. By reshaping parts in a heated press, additional energy is consumed. It needs to be verified if the carbon footprint of a refurbished organic sheet after a second life is lower than the carbon footprint of a virgin organic sheet. Therefore, a life cycle assessment needs to be performed to prove if a refurbishment can contribute to a circular economy approach of organic sheets. Nevertheless, this paper has shown that two-dimensional specimen organic sheets can be refurbished and used again with a low degradation of mechanical properties for high cycle fatigue.

Acknowledgements. The German Federal Ministry of Research and Education BMBF funded this research under funding number: L1FHG42421. Additionally, the authors thank the Fraunhofer-Gesellschaft für angewandte Forschung e. V., the Fraunhofer-Institute for Machine Tools and Forming Technology IWU and the Open Hybrid LabFactory e. V. for their support.

References

1. Sauer, M.: Composites-Marktbericht 2019: Der globale CF- und CC-Markt 2019: Marktentwicklungen, Trends, Ausblicke und Herausfordeungen
2. Witten, E., Mathes, V.: Der Markt für Glasfaserverstärkete Kunststoffe (GFK) 2020: Marktentwicklungen, Trends, Ausblicke und Herausforderungen. Industrievereinigung Verstäkte Kunststoffe (2020). Accessed 28 Jan 2022. www.avk-tv.de/files/20201111_avk_marktbericht_2020.pdf
3. Das, S.: Life cycle assessment of carbon fiber-reinforced polymer composites. Int. J. Life Cycle Assess. **16**, 268–282 (2011). https://doi.org/10.1007/s11367-011-0264-z
4. Ghanbari, A., Seyedin, S., Nofar, M., Ameli, A.: Mechanical properties and foaming behavior of polypropylene/elastomer/recycled carbon fiber composites. Polym. Compos. **42**(7), 3482–3492 (2021). https://doi.org/10.1002/pc.26073
5. Asmatulu, E., Twomey, J., Overcash, M.: Recycling of fiber-reinforced composites and direct structural composite recycling concept. J. Compos. Mater. **48**(5), 593–608 (2014). https://doi.org/10.1177/0021998313476325
6. Freeden, J. v., Erb, J., Schleifenbaum, M.: Separating layer recycling strategy for continuous fiber reinforced thermo-sets based on thermally expanding particles. Polymer Composites, pp. 1–13 (2022). https://doi.org/10.1002/pc.26505

7. Kiss, P., Stadlbauer, W., Burgstaller, C., Fehringer, S., Haeuserer, F., Archodoulaki, V.-M.: In-house recycling of carbon- and glass fibre reinforced thermoplastic composite laminate waste into high performance sheet materials, Composites Part A: Applied Science and Manufacturing **139** (2020). https://doi.org/10.1016/j.compositesa.2020.106110
8. Kunststoffe – Bestimmung der Zugeigenschaften – Teil 4: Prüfbedingungen für isotrop und anisotrop faserverstärkte Kunststoffverbundwerkstoffe, DIN EN ISO 527-4 (1997)
9. Kunststoffe – Bestimmung der Zugeigenschaften – Teil 4: Prüfbedingungen für isotrop und anisotrop faserverstärkte Kunststoffverbundwerkstoffe, DIN EN ISO 527-4-draft (2020)

Manufacturing of Lightweight Parts by Sandwich Foam Injection Moulding using Recycled Thermoplastics

Annerose Hüttl[1(✉)], Mathias Kliem[1], and Ralf Utescheny[2]

[1] Kunststoff-Zentrum in Leipzig gGmbH,
Erich-Zeigner-Allee 44, 04229, Leipzig, Germany
{huettl,kliem}@kuz-leipzig.de
[2] Leichtbau Zentrum Lupburg, Koller Kunststofftechnik GmbH,
Degerndorf G 2, 92331, Lupburg, Germany
r.utescheny@koller-gruppe.de

Abstract. Reinforced, high-performance thermoplastics may have moduli of elasticity of up to 40 GPa, depending on the filler content. The filling materials can thereby consist of secondary raw materials such as recycled carbon fibers. By combining these highly stiff materials with unreinforced but material-compatible and foamable thermoplastics, excellent and complex lightweight sandwich structures can be realized in a one-shot process using the sandwich foam injection moulding technology. The core and top layers of these structures are usually made of two different components (2-C). The process is both reproducible and cost-efficient for high production volumes. The foaming process is carried out either by adding a chemical blowing agent or an inert gas at supercritical conditions. The resulting composite consists of a short-fiber reinforced high-modulus material in the outer layer and an unreinforced foamed material in the core. Due to the low requirements on performance, odour as well as surface properties, recyclates are particularly well suited for the use as core material. Three sources of recyclates are presented in the paper, which have been the scope of the specific studies: A) Polyolefin mixed fraction from household waste, B) PET (polyethylene terephthalate) waste flakes, and C) polyamide waste. Depending on the source mentioned, dedicated lightweight structures with specific properties may be produced—with applications in particular market segments. A suitable example in this context is the sandwich foam injection moulded functional support structure for the electro-mobility automotive industry with a design suitable for foam injection moulding. Moreover, a significantly higher specific stiffness was achieved with this semi-structural component compared to an existing reference structure.

Keywords: Foam injection moulding · Recyclates · Sandwich structure · Lightweight design

K. Dröder and T. Vietor (Eds.): *Future Automotive Production Conference 2022*, Zukunftstechnologien für den multifunktionalen Leichtbau, pp. 30–43, 2023.
https://doi.org/10.1007/978-3-658-39928-3_3

1 Introduction

The German Federal Ministry for Economic Affairs and Energy designates lightweight design as a key technology that enables industry to manufacture lighter and more cost-effective products with improved resource efficiency [1]. In the future, automotive mobility will be characterized by the development of intelligent and at the same time sustainable lightweight solutions [2, 3]. Thermoplastic foam injection moulding (FIM) [4] is a promising and future-proof process for implementing lightweight solutions. With its application, particularly lightweight and yet robust, strong and durable components may be manufactured [5, p. 4]. Political and legal frameworks aim to encourage manufacturers to intensify the use of recyclates in order to significantly increase their market share [6, 7].

The Kunststoff-Zentrum in Leipzig gGmbH (KUZ) uses the FIM process to manufacture sandwich components from recyclate. This is intended to contribute to the "European Strategy for Plastics in the Circular Economy" as part of the European Green Deal [8]. In accordance to this strategy, the use of recyclates is to be increased from 4 to 10 million tons by 2025 [6]. However, higher recycling rates can only be achieved if high-quality recyclates are made widely available in sufficient quantities. The availability of sorted post-industrial waste on the European market tends to become increasingly limited, as the market demand for higher-quality recyclates, which may be used in technical applications, is steadily increasing [9, 10]. The KUZ therefore uses waste from post-consumer recycling (PCR) as a source of raw materials. Table 1 exemplarily presents foamed sandwich components from three different waste streams (see Table 1).

Table 1. Types of foamed recycled material, sources and possible applications.

	Type of post-consumer recycling (PCR)	Scope
1	PE/PP—mixed waste from the household waste fraction	Automotive interior and construction industry
2	Polyethylene terephthalate (rPET) from bottle waste of the household waste fraction	Electrical and electronic applications, thermally insulating housings, connectors, insulating containers, sound insulation, sports equipment, maritime floating bodies
3	Polyamide (PA) from fiber waste, ropes and fishing nets	Automotive applications and mechanical engineering with increased heat load, e.g. housing and module cover for Li-ion batteries

2 Material Structure According to Sandwich Principle

The sandwich principle leads to a reduction in weight with comparable bending stiffness, so that the lightweight design effect clearly has an impact. At the KUZ, component geometry, materials and manufacturing processes are combined in such a way that the sandwich principle can be applied effectively while reducing component costs. Figure 1 schematically shows a bending loaded structural part under load F. Compressive stresses occur on the side where the force is applied (marked in blue), whereas tensile stresses occur on the opposite side (marked in green). In the neutral axis, the normal stresses decrease to zero, whereas the normal stresses in the face layers become maximum. According to Parallel axis theorem, the load-bearing capacity increases as the distance between the face sheets and the neutral axis is increased. The sandwich core essentially acts as a spacer and must absorb the shear stresses that become maximum in the core. Application of the sandwich principle means that high-strength materials are ideally placed far away from the neutral axis and lightweight materials are placed close to it.

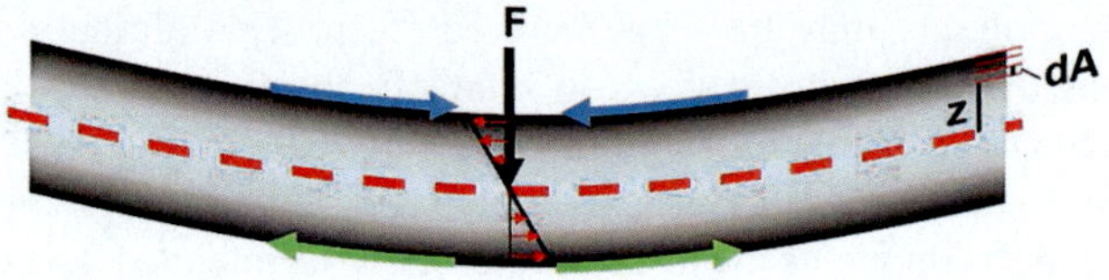

Fig. 1. Schematic representation of the sandwich principle

The high strength and stiffness of the short-fiber-reinforced material of the outer layer contributes significantly to the bending stiffness, while the core made of lightweight but strong foam enhances the weight reduction. The 2-component FIM manufacturing process allows the use of different thermoplastics in the outer and core layers. The outer layer is made of a reinforced plastic that provides both the high-quality visual appearance and the structural function. The core contains a lightweight foam made of unreinforced shear-resistant thermoplastic. In the most basic case, the foam can be achieved using a chemical blowing agent. An option for large-scale production involves physical direct gassing of the melt using nitrogen.

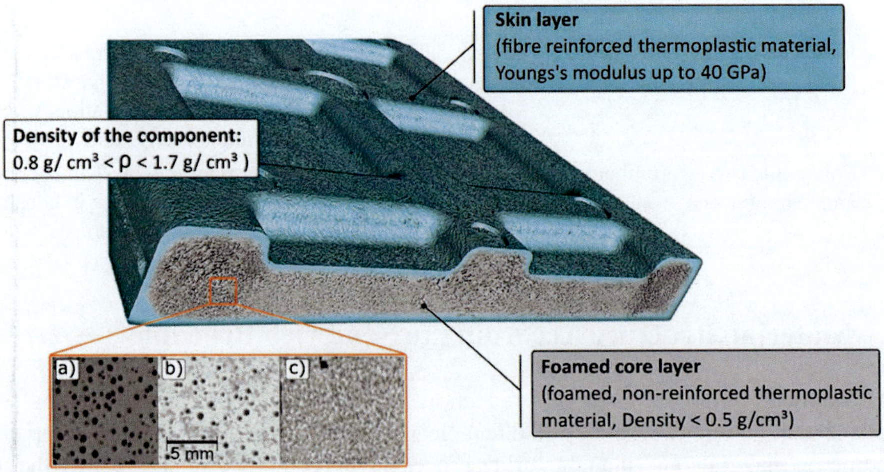

Fig. 2. Exemplary principle of sandwich-like material structure | a), b), c): Improvement of the foam quality of the recycled material in the core through upstream processing step.

Figure 2 shows a cross-section of a test specimen produced in the 2-C FIM process using the imaging computed tomography technology available at the KUZ. The test specimen serves as a demonstrator for two-dimensional, shell-shaped industrial applications with a structural function. Hereby, the higher weight-specific bending stiffness compared to a part made of compact material may be well illustrated.

3 Manufacturing Process and Machine Requirements

For the processing of two different thermoplastic materials, a 2-C injection moulding machine is required. The two separately adjustable plasticizing units allow the injection moulding parameters to be set in order to fulfill the requirements of the respective part (e.g. skin/core layer distribution). For the specific control of the skin and core melt flow, a commercially available 2-C sandwich spacer plate is used between the mould and the plasticizing cylinder. Its working principle is shown schematically in Fig. 3. During 2-C sandwich injection moulding, the cavity is partially filled with the skin material (1st component) in the first step, and a thin, solidified skin layer is formed in contact with the cold mould wall. The immediately following foamed core material (2nd component) further displaces the remaining plastic core of the skin material until the cavity is completely filled. A needle valve nozzle prevents the gas-loaded melt from leaking. Foam is formed only after the pressure in the mould cavity has dropped.

When using a chemical blowing agent, which is usually added to the core layer material as a masterbatch, a 2-C injection moulding machine with standard plasticizing units may be used. Physical direct gassing with nitrogen is more suitable for higher volumes due to the higher machine-related investment costs. The process also leads to lower foam densities due to higher foam pressures and thus provides increased lightweight design potential [11, 12].

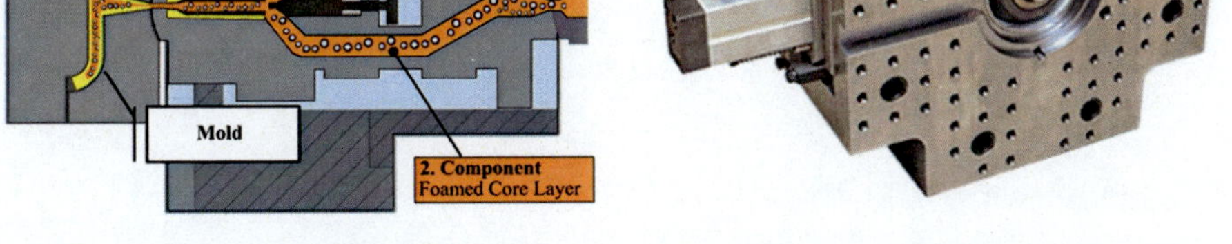

Fig. 3. Left: Cross-sectional sketch showing the working principle of the sandwich spacer plate, the melt flow of component 1 (skin material) in the vertical plasticizing unit and component 2 (core material) in the horizontal plasticizing unit. Right: Top view of sandwich spacer plate from A&E Produktionstechnik GmbH, Dresden.

Figure 4 provides a schematic representation of the process combination 2-C sandwich injection moulding with physical direct gassing. The experimental tests are carried out at the technical lab of the KUZ on a 2-C injection moulding machine (type: Combimould HM-MK 180/525H/350V with Cellmould® unit for physical foam generation by WITTMANN Technology GmbH).

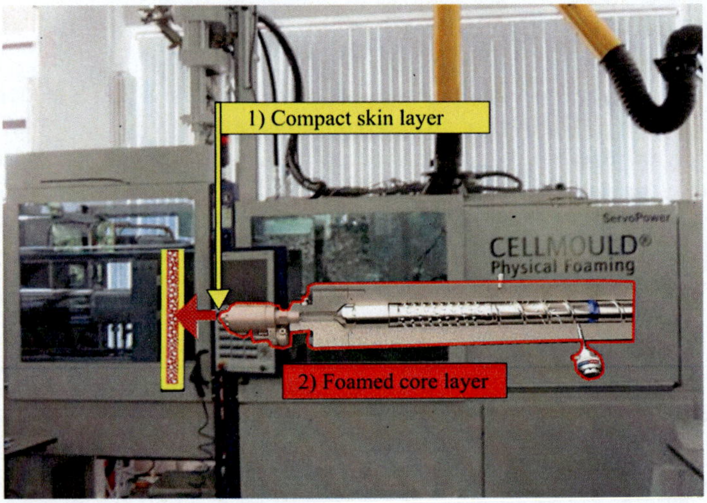

Fig. 4. Schematic representation of the process combination

4 Material Preparation as a Prerequisite for Foaming Post Consumer Recycled (Pcr) Material

The challenge in foaming PCR materials is the thermal-oxidative molecular degradation caused by environmental degradation during the life cycle as well as renewed thermal stresses as part of the recycling process. The resulting melt instability often leads to rupture of the bubble webs during the foaming process. This in turn leads to bubble coagulation, which may eventually impair the mechanical

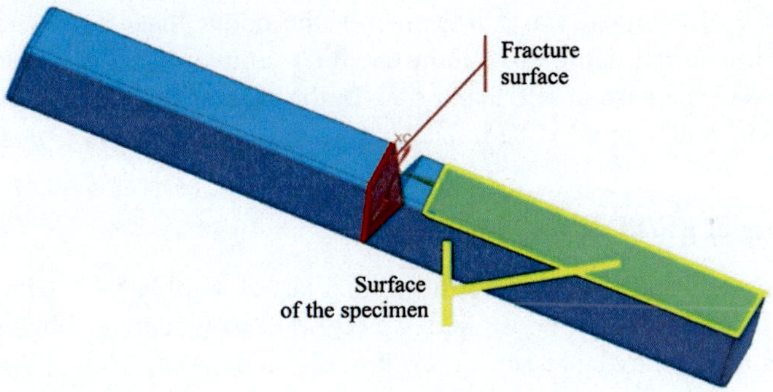

Fig. 5. Notched test specimen

properties of the structural parts. Since gas is prevented from leaking through the parting line within a 2-C sandwich mould due to the closed shell of skin material, a non-uniform foam structure often occurs. A fine-cell foam structure develops near the sprue, while large bubbles appear at the flow path end. In view of the negative effects on the mechanical properties and dimensional stability as well as shrinkage and warpage, the bubble size distribution of the core layer should be as homogeneous as possible. The necessary processing step of regranulation for reasons of flowability and feedability into the injection unit may be used for material preparation by means of additives or reactive compounding.

5 Test Series for Recipe Optimization

A special notched test specimen, shown in Fig. 5, is used for an initial formulation adjustment of the material composition with respect to impact strength, subjective odour characterisation, foam structure and moulded part surface. Using the test specimen shown above, test series for improving properties are investigated (see Tables 2 and 3) and material modifications are deduced. By adding additives in a preceding compounding step, an improvement of the properties to those of the virgin material is intended. The dimensions of the specimen used for the three test series differ significantly from those of the usual specimen for the notched bar impact test according to DIN EN 179-1. Based on the specific specimen, a rapid evaluation of bubble size distributions across the test specimen cross-section may be obtained using an automated algorithm, developed at KUZ.

6 Preparation Steps to Match the Properties of Fresh Material

In the following chapter, three different PCR materials rPP/rPE, rPET as well as rPA are examined with regard to their respective improvements in foam quality. While in the case of rPP/rPE mixed waste it is mainly the odour that is a decisive exclusion criterion for use in the automotive industry, for example, the focus is mainly on the foam structure in the case of rPET and rPA. In the case of rPET, the degree of embrittlement, demonstrated here by the impact strength, is also relevant.

6.1 Foaming of PE/PP Mixed Waste

The PCR regranulate Dipolen S® from mtm Plastics GmbH in Niedergebra, which originates from large-scale household waste processing and is obtained from the PE-PP mixed fraction by large-scale recycling plants, is used.

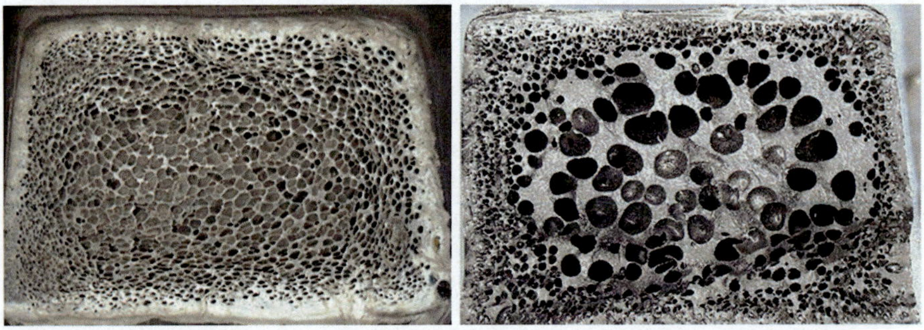

Fig. 6. Cross-sections of the notched test specimen made from the Dipolen S® regrind I without the addition of additives (left), with the addition of additives (right)

A sound and homogeneous foam structure can be achieved even without the addition of additives (see Fig. 6, left). Figure 6 (right) shows an integral foam structure altered by the compound modification as shown in Table 2. The incorporated additives (Additive 1–4) lead to improved foam morphology due to the increase in melt stability. In both cases, a weight reduction of about 30% is achieved. A critical concern with PCR from the household waste fraction is the strong odour, which has so far widely prevented its use in industrial applications. A reduction in odour and emissions can be achieved by adapting the formulation in a preceding compounding step. By using an entraining agent, oligomers are removed from the melt, so that the odour is significantly reduced compared to the reference material (see Table 2). However, this is still not sufficient for an application in the automotive interior sector for reasons of emissions and the part surface, so that encapsulation by processing as a core layer in 2-C sandwich moulding could be an appropriate approach for this purpose.

Table 2. Odour reduction

Compound	1 (Reference)	2	3	4
Dipolen S	100%	97,7%	95,7%	92,7%
Entraining agent		2%	2%	4%
Additive 1			1%	1%
Additive 2			1%	2%
Additive 3		0,1%	0,1%	0,1%
Additive 4		0,2%	0,2%	0,2%
Odour (Grade, subjective, without norm, 5 – intense, 1- hardly none)	5	3,5	3,5	2

The results presented in Table 2 for the assessment of the odour level are based on a purely subjective perception. No odour testers specialised in this field were involved. The values should therefore only be regarded as guidelines and would have to be measured again in a qualified manner as the project progresses. Irrespective of the above, it can be stated that the recycling odour typical of PP/PE mixed waste may be reduced by about half when entraining agents and additives are incorporated. The specific type designation of the additives for the odour reduction of PE/PP mixed waste and for molecular chain attachment for rPET (see Sect. 6.2) are subject to confidentiality due to a new development and can therefore not be published.

6.2 Foaming of rPET from Bottle Waste

Figure 7 on the left shows the grinded flakes of MultiPet GmbH Bernburg as delivered, after re-granulation without additives (Fig. 7, center) and as compound processed with additives (Fig. 7, right).

Fig. 7. Left: rPET flakes as supplied, Center: After re-granulation without additives, Right: Processed with additives

Without the addition of additives, the material with a light brownish discoloration shows a residual contamination by very fine particles (see Fig. 7, center). In addition, the flakes are not of sufficient quality for foaming due to insufficient melt strength. Figure 7 right shows the prepared foamable granulate.

In order to be able to feed the light flakes into the injection moulding machine without difficulties, the necessary regranulation is used at the same time as a reconditioning step for the addition of additives to improve the part surface and impact strength. The additives are incorporated using a co-rotating ZE25Ax47D-UTXi-UG twin-screw extruder (screw: 25 mm, L/D: 47, L: 1175 mm) from KraussMaffei Berstorff. The differences in weight reduction, impact strength and foam structure between the regrind from rPET bottle waste made from pure flakes and the prepared compound are shown in Table 3.

Table 3. Properties of foamed rPET after preparation

	Regrind from rPET bottle waste	Reprocessed compound
	Chemically foamed	
Weight reduction (%)	8	28
Impact strength (kJ/m^2)	13	35

Figure 8 Left shows the cross-section of a foamed sample made of rPET in its initial state without the addition of additives. Individual voids, an uneven component surface and a clear yellow discoloration as a result of degradation mechanisms can be seen. The cross-section of a foamed test specimen produced using an additivated compound is shown in Fig. 8, right. A very fine-cell foam structure is formed as shown in the detail magnification of an SEM image (see Fig. 8 below). The bubble size is in the range of about 100 µm. The surface is smooth, the visual impression greatly improved and no yellow discoloration occurs.

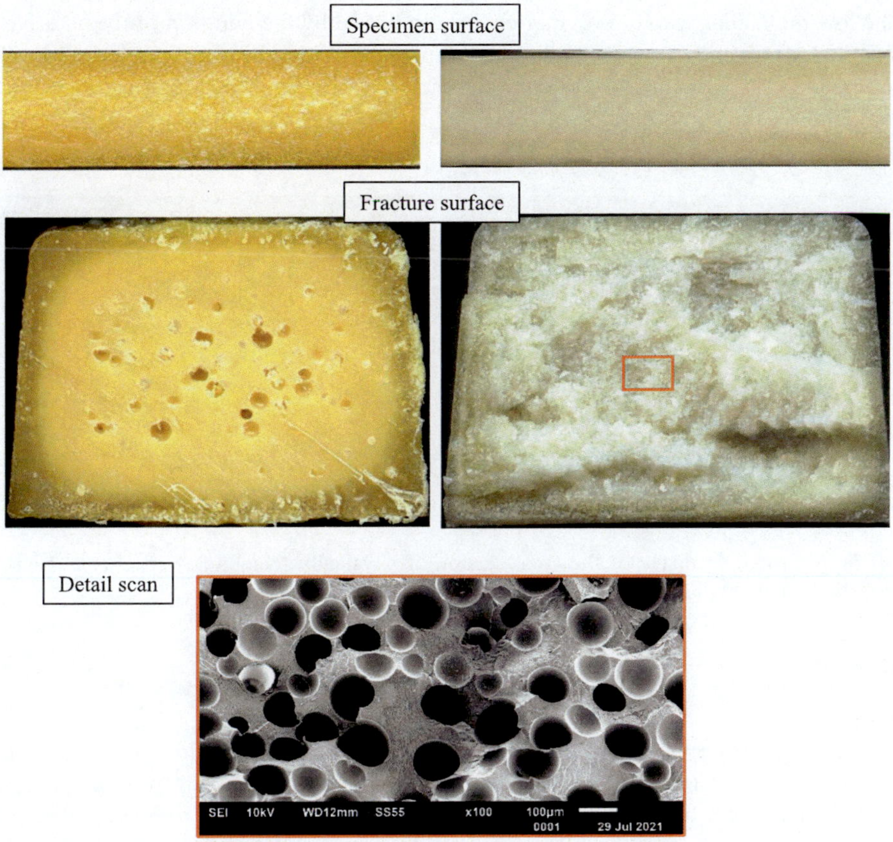

Fig. 8. Left: PET recycled foam before reprocessing with yellowish discoloration, uneven specimen surface and insufficient foam structure due to molecular degradation reaction, Right: After preparation with smooth specimen surface, without discoloration and with improved foam structure, Bottom: Detail picture (SEM) of the compound with additives

6.3 Foaming of PA6 from Fiber Waste, Ropes and Fishing Nets

Due to thermal-oxidative molecular degradation reactions, this recycled material also exhibits insufficient melt strength for foaming in its as-received state. The rheological characteristics such as the melt flow rate MFR (275 °C/0.325 g) of 10 g/10min and the viscosity number of 135 ml/g indicate a reduced molecular weight and thus poor foamability. During both chemical and physical foaming, large voids with a diameter of approx. 1 to 3 mm repeatedly occur, which predominantly concentrate at the end of the flow path. By adding 1% of the Chain Extender MB AR/PBT-30, Trigon Chemie GmbH [13], the foam structure could be decisively improved (see Fig. 9). The chain extender is available in the form of a masterbatch and consists of a preparation of benzene-1,2:4,5-tetracarboxylic acid dianhydride, pyromellitic acid dianhydride, benzene-1,2,4-tricarboxylic acid 1,2-anhydride and trimellitic acid anhydride.

The low molecular weight polymer chains are linked by the addition of the additive to form longer chain polymers while maintaining the original linear structure [14]. The compounding step with reactive compounding is carried out on a twin-screw extruder analogous to the compounding of rPET.

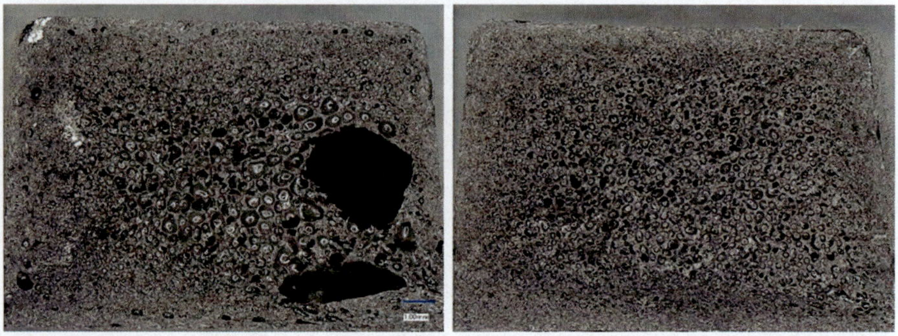

Fig. 9. Microscopic image of the cross-section of physically foamed test specimens made of PA6 (Left: Without chain extender, Right: With chain extender)

7 Production of Sandwich Component Function Carrier

The findings described in Chap. 6 on improving the foam quality of PCR materials of one-dimensional test specimens can be applied to structurally more complex components, such as the sandwich component functional carrier shown in Fig. 10. Additives are used to improve the foam quality depending on the core material used (rPP, rPET) in the same way as in the previous test series. In cooperation with the Koller Leichtbau-Zentrum Lupburg, a R&D project has resulted in the development of a topology-optimized semi-structural lightweight part (see Fig. 10). It is based on a bionic design and is manufactured using a 2-C sandwich injection moulding process. In order to be able to use the one-shot sandwich process economically for largescale structural and visible components in high-volume series production, a multiple connection with hot runners and cascading filling is a necessary requirement. To this end, Koller has contributed a patent-pending mould development to the project.

Fig. 10. Topology-optimized semi-structural lightweight part using a 2-C sandwich injection moulding process

Two types of PCR material are processed in the core layer. The material compositions are:

- Material cluster A
  Skin layer: virgin material PP-LGF40 Fibremod GB 402 HP, Borealis AG.
  Core layer: Processed regranulate of PE/PP—mixed waste of the domestic waste fraction
- Material cluster B
  Skin layer: virgin material PC/PET-T15 macroblend VT 235 M, Covestro AG
  Core layer: rPET from processed regranulate from bottle waste of the household waste fraction

8 Possibilities of Industrial Applications and Conclusion

The production of lightweight structures in 2-C sandwich injection moulding is well suited for large-scale production due to the one-shot technique and thus provides the potential for cost-effective lightweight engineering. Ribbed, flat, plate-like or shell-shaped components with a load-bearing function are predestined. For the bending load case, the sandwich effect results in weight reduction while retaining the mechanical properties. For rotating and/or fast-moving components, energy savings can be achieved by reducing the moving masses.

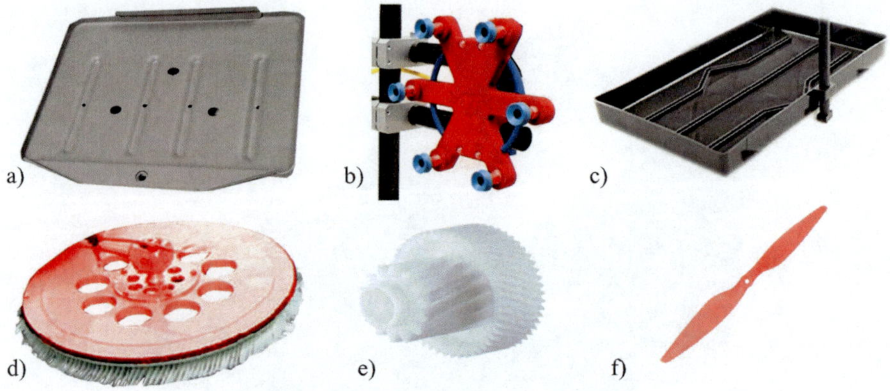

Fig. 11. Possibilities of industrial applications, such as a) cover elements, b) gripper elements, c) housing components, d) fast rotating components, e) machine elements and f) rotor components

Some examples within the field of automotive applications include: Front end carrier, pillar trim, sill trim, battery carrier, hood for cylinder head, hood compartment cover, step trim, seat pan, trunk floor, rear spoiler. Other technical applications are: High strength planar housings, structures for robot heads, moving closing mechanisms, fast rotating machine elements (see Fig. 11). Applications for foamed rPET are mainly in the field of electrical and electronics. Recycled PA foam may be used in components with increased thermal or mechanical stress.

Based on the presented test studies with the three PCR compounds rPP, rPET and rPA, it is demonstrated that the foam quality, the foam morphology and the impact strength can be substantially improved by a systematic preparation of the materials through the addition of suitable additives. Depending on the material, various mechanism of actions are used, such as increasing the melt stability by adding chain extenders or reducing odour due to the incorporation of entraining agents. For the material PP/PE mixed waste it can be stated that the odour level may be decisively reduced by the incorporation of various additives and entraining agents. A significant improvement of the foam morphology as well as a reduction of degradation effects may be achieved by adding various additives for rPET. For rPA, a significantly improved foam morphology may be obtained by means of the application of chain extenders. In the course of the experiments, the additives are added by means of a preceding compounding process step in order to ensure an effective incorporation of the additives. In the further process of the project, the material properties are to be investigated in comparison with the respective virgin material.

Acknowledgements. Thanks to the German Federal Ministry for Economic Affairs and Energy for the financial support of the project 'RecySchaum' (49MF200019). The authors are also indebted to mtm Plastics GmbH and MultiPet GmbH Bernburg, providing various PCR materials.

References

1. Bundesministerium für Wirtschaft und Energie: Artikel Schlüsseltechnologien – Leichtbau, Ausgabe (2020)
2. Horst E.: Leichtbau in der Fahrzeugtechnik (2. ed.), p. 178 (2017)
3. Pothen, F., Growitsch, C., Engelhardt, J.: Fraunhofer IMWS, Stiftung Arbeit und Umwelt der IG BCEDie Automobilindustrie im Wandel Beschäftigungspolitische Implikationen des Automobilsektors für die chemische, die gummi- und kunststoffverarbeitende Industrie, Oktober 2019, p. 34
4. Gomez-Monterde, J., Hain, J., LI. M.: Maspoch: Characterization of microcellular plastics for weight reduchtion in automotive interior parts, Kunststoffe im Automobilbau, VDI Verlag GmbH, Düsseldorf, p. 167 (2017)
5. Braun, J., Gahleitner, M., Grestenberger, G., Niedersüß, P., Tranninger C.: Wachsen durch technische Spezialitäten, Special K2019, Weltmarkt der Kunststoffe, Carl-Hanser-Verlag München, Kunststoffe 10/2019
6. Caudet S.: Circular Plastics Alliance—100+ signatories commit to use 10 million tons of recycled plastic in new products by, 2025 European Commission—Press release. Brussels 20.09.2019
7. European Commission Press Release: Commission launches Circular Plastics Alliance to foster the market of recycled plastics in Europe, Brussels, 11.12.2018
8. Suica D.: Keynote speech at Jutarnji List World of Energy Conference, European Commission, Brussels, 26.01.2022
9. EUWID Europäischer Wirtschaftsdienst GmbH, Recycling and Waste Management, Further price hikes expected on German waste plastics market, 12.01.2022
10. Stachura M.: Five rules for plastic processors using recycled materials sustainably, Sustainable Plastics, p. 1 (2021)
11. Hüttl A.: Cell-2K—Herstellung von Leichtbauteilen durch Verfahrenskombination Physikalisches Schäumen – 2-Komponenten-Sandwich-Spritzguss, Reg.-Nr. MF160109, Projektlaufzeit 11/2016–06/2019
12. Hüttl, A., Kliem, M.: Fibre reinforced lightweight structures using the sandwich injection moulding technology and the physical foaming process, Presentation at Merge, 4th IMTC 2019, 19.09.2019
13. Technisches Datenblatt Chain Extender MB AR/PBT-30, Trigon Chemie GmbH, Schlüchtern, 13.01.2020
14. Kliem, M., Hüttl, A.: Upcycling von Polyamid Mahlgut—Verbesserung der Schäume-Qualität, Praxisforum Kunststoffrezyklate 2020 Hanser, Darmstadt, 04.03.2020

Functional Structures

Innovative Design Concept for the Safety of Battery Housing

Claudia Drebenstedt[1](✉), Marcus Knobloch[2], David Löpitz[2], Elisa Ruth Bader[2], and Patryk Nossol[2]

[1] Department Lightweight Design, Fraunhofer Institute for Machine Tools and Forming Technology IWU, Reichenhainer Strasse 88, 09126 Chemnitz, Germany
claudia.drebenstedt@iwu.fraunhofer.de

[2] Department Lightweight Design, Fraunhofer Institute for Machine Tools and Forming Technology IWU, Chemnitz, Germany
{marcus.knobloch,david.loepitz,elisa.bader,
patryk.nossol}@iwu.fraunhofer.de

Abstract. The automotive production is a very energy- and resource-intensive industry. For that reason, it is important to rethink many structures from the state of the art with the aim of generating a more environment-friendly future, for example by means of circular economy. In order to achieve this aim, a vehicle platform with a long service lifetime is required. The design of future parts should be much more modular to have the possibility of replacing parts easily, if needed. This basic approach is followed in a research project regarding the design of battery housings for electrical vehicles. One of the main research objectives in this project is to reduce the mass while increasing the lifetime and the crash safety of a modular battery housing for electric cars. Especially for the side crash the sill structure is essential. Therefore, an innovative modular concept was developed, which uses multiaxial fiber-reinforced crash tubes. To manufacture these tubes in a close to series production, the pultrusion process was used and technically adapted. This paper presents the findings of the design, process layup, manufacturing steps, and results of an adapted pultrusion process to produce load adjusted crash tubes made of fiber-reinforced plastics (FRP).

Keywords: Lightweight design · Crash tube · Modular design · Fiber-reinforced plastics · Pultrusion

1 Introduction

Until now, the automotive industry has been organized predominantly in linear processes. Although vehicles are mostly made of parts that could be reused, these are hardly ever used for the design of new vehicles. Rather, at the end of the vehicle's

K. Dröder and T. Vietor (Eds.): *Future Automotive Production Conference 2022*, Zukunftstechnologien für den multifunktionalen Leichtbau, pp. 47–58, 2023.
https://doi.org/10.1007/978-3-658-39928-3_4

lifetime, the parts and materials are often downcycled to less demanding applications [1]. Ecologically advantageous is a reuse at the component or module level. Approaches to such a circular economy can be seen in the supply of spare parts [2, 3] but the multiple reuses of large, complex vehicle modules or vehicle structures is not yet known. The main approach of the research project aims at a modular, long-life vehicle platform (see Fig. 1), which means that parts or entire modules can be used more than one service lifetime [4].

- Business model development
- Environmental accounting
- Concepts for modular design
- Battery housing
- Crash concept: design of sill
- Functional integrated suspension element

Fig. 1. Circular open-source design kit

Therefore, a higher lightweight design aspect can be used, which is usually associated with a higher production effort and thus higher production and materials costs. The CO_2 reduction and cost efficiency are achieved over the longer lifetime of the platform. Due to its modular built-up, it is possible to easily replace damaged parts. One of the main focuses in the research project for the Fraunhofer IWU is the increase of the crash safety with fiber-reinforced crash tubes in the side sill of an innovative modular car concept. State of the art for the side-impact crash safety is mainly the usage of complex structures made of metal sheets or extruded multi-chamber profiles [5]. In case of the usage of steel, corrosion leads to a reduction in safety. Repair or replacement in case of a (minor) crash is also often complicated due to the integration into the vehicle structure, for example by welding seams.

Fiber-reinforced solutions have so far mainly served to provide structural cover for design reasons without making any real contribution to crash safety. Crash relevant sill structures made of FRP are mainly still in the field of research, since the fiber orientation plays a decisive role in force or load absorption. The state of the art shows many studies of the crash behavior of a tube made of FRP, its testing methods and failure modes [6–8]. These studies show that a multiaxial fiber structure is suitable for crash applications providing a good energy absorption. In [9] is shown that a fiber orientation consisting of $\pm 30°$ is optimal to provide a maximum energy absorption. Due to these tough requirements, many approaches are currently being pursued using the fiber winding or resin transfer molding (RTM). However, these processes are too time-consuming and cost-intensive for use in a large-scale production. One of the few manufacturing processes suitable for large-scale production of fiber-reinforced tubes is the pultrusion process. The advantage of using pultrusion as manufacturing process for those tubes is discussed in several papers [10, 11]. The state of the art shows that tubes with an unidirectional 0° orientation can be produced by pultrusion with a speed of more than 1 m/s in a very high quality. However, the unidirectional fiber orientation is not optimal for crash applications since the energy absorption already decreases significantly at a deviation of the loading direction of a few degrees from the 0° axis.

In order to exploit the potential shown in conjunction with improved series suitability, the crash tubes focused on in this work are manufactured using a pultrusion process with a complex scrim structure based on the results of finite element simulations [12].

2 Design and Process Setup

The following part describes the iterative cycle of design.

2.1 Design

The focus of this paper is on the work of the Fraunhofer IWU, which concentrates on the battery housing and the crash safety for the side impact [4]. The aim was to design a modular, lightweight vehicle platform, so that damaged parts could be exchanged easily. Previous internal studies showed that there is a high potential for carbon fiber reinforced plastics (CFRP) crash tubes regarding lightweight and the energy consumption of a crash and therefore the security of the battery and the occupants of the vehicle [9, 13].

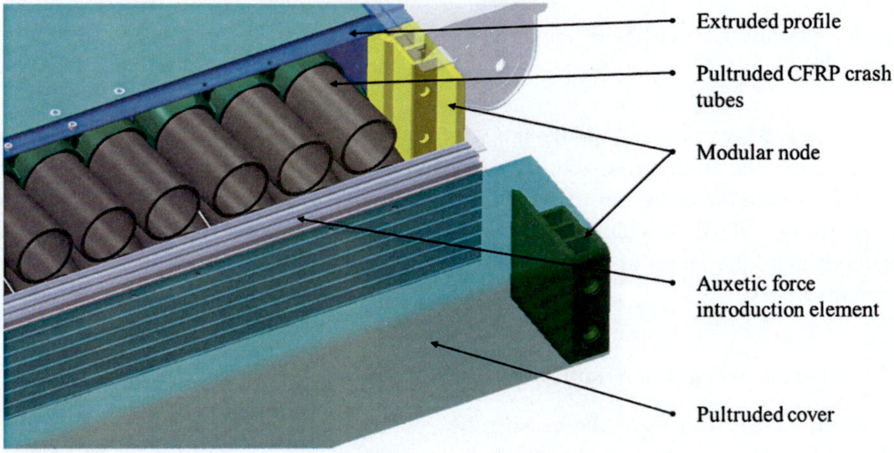

Fig. 2. Modular built-up of the battery casing with sill and crash tubes

Based on these investigations, the high potential of these crash tubes for the use in a new type of sill structure was presented in [12].The design of the sill structure is shown in Fig. 2. The side is equipped with two rows of the crash tubes which are inclined with + 5° to the y-axis on top and −5° in the row below (see Fig. 3). This special angular orientation of the crash tubes was determined on the base of previous experimental studies. This means that impact forces acting at different angles can also be better absorbed. An extruded buffer element (Fig. 2, light grey) is placed above the crash tubes. It is designed as an auxetic structure, so the acting force is more evenly distributed and directed into the crash tubes. This special behavior for auxetic buffer elements was proven experimentally and computationally [12, 13]. Above these structures is a pultruded cover (Fig. 2, transparent green). An adapter holds three of these crash tubes. This adapter has the function to hold the crash tubes in place, so it will fail according in the type of crushing instead of being pushed away.

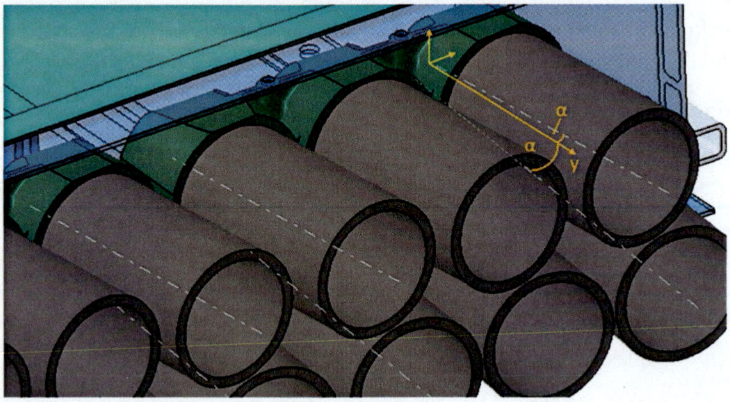

Fig. 3. Adapters with crash tubes inclined at 5° to the y-axis

As an iterative process, the design was the base for the mechanical simulation and the results of the simulation were used to adapt the design.

To evaluate the design of the crash tubes and the sill structure under the restriction of the EURO NCAP standard [14], FEM-simulations were conducted. In this way, the diameters and the layup of the tubes including the wall thickness of the adapters were examined [12].

2.2 Process Setup and Manufacturing Concepts

After the design freeze of the crash tubes, the design of the pultrusion die and the guiding panels started. For the crash tube demonstrator, a pultrusion die with the length of 400 mm was designed and manufactured (Fig. 4). The core contains three additional heating cartridges, which ensure that the profile is completely cured despite a relatively high wall thickness of 5.5 mm.

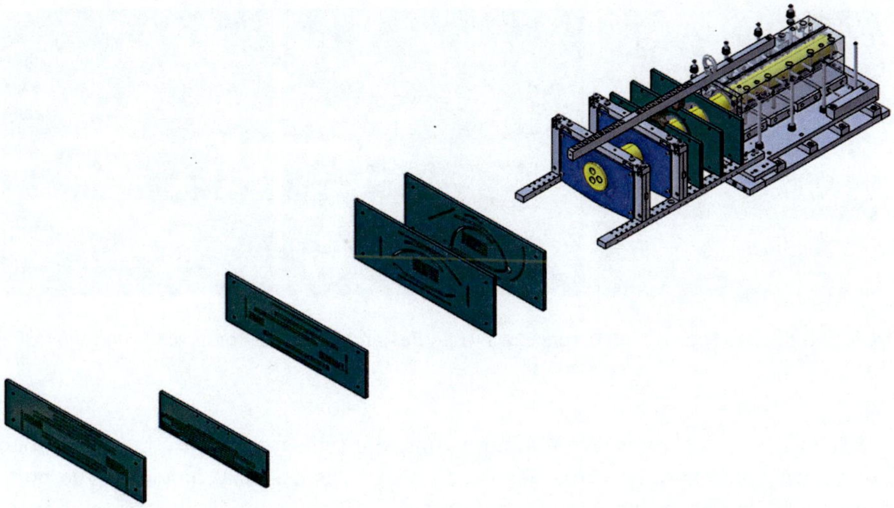

Fig. 4. CAD-Model of the die, die carrier, floating core, and guiding panels

A total of 10 guiding panels are required to guide the semi-finished products in an orderly manner through the open resin bath and to preform the tube geometry before entering the die. Two of the guiding panels are used to hold the core (Fig. 4, dark blue), which "floats" in the die to create a hollow profile. The second guiding panel from the left side in Fig. 4 leads the fibers and the semi-finished products through the resin bath. When using fabrics or scrims in complex shapes, wrinkling must be avoided. This means the guiding panel slots height must be a compromise between too high (allowing the fabric to wrinkle) and too low (damaging the multiaxial fiber structure).

The pultrusion of tubes made of rovings in general is state of the art. The challenge in this project was to manufacture the tubes with the required complex layer structure. According to the experimental results for suitable layer orientation for crash tubes [9] and the computations [12], a fiber orientation of $\pm 30°$ at the main portion is desired to achieve best mechanical properties, especially a high absorption of energy in the case of a crash to secure passengers and battery. To achieve the desired properties, as few seams as possible are optimal. Therefore, two general concepts were considered, shown in Fig. 5.

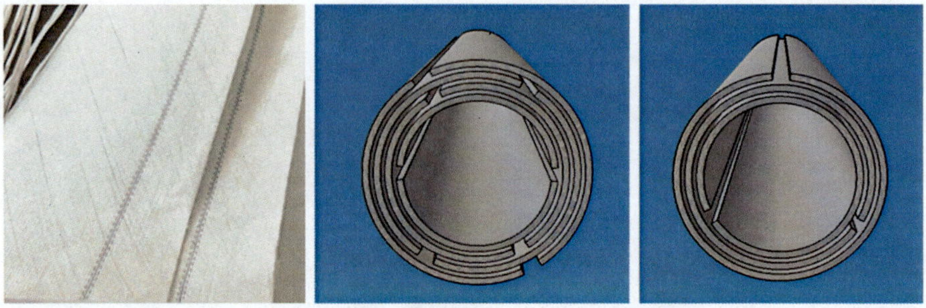

Fig. 5. Multiaxial scrims (left), concept 1 (middle) and 2 (right) for possible lay-ups of the scrim

In concept 1, the layers are split into two textile semi-finished products. In concept 2, one layer consists only of one textile. Concept 1 is easier to handle in the pultrusion process, but an accurate positioning of the edges of the scrims of one layer to another is difficult. Concept 2 has fewer seams, but the supply is much more complicated, resulting in a special design of the combined fiber guide and core bearing, which leads to a higher probability of wrinkling. The seam positions are arranged at different angles around the profile to balance their influence on the mechanical properties of the profile.

While using rovings, a fiber volume content (FVC) of 60 to 70% is usual. The achievable FVC with scrims is below this and considerable depending on the scrim used. To test those limitations, the first layup was designed to have a very high portion of scrims: an inner and outer layer of rovings and 4 layers consisting of two scrims per slot in the guiding panels.

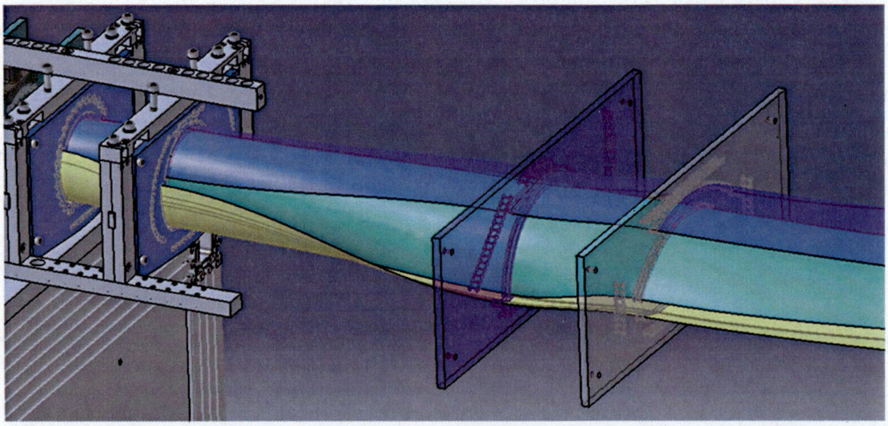

Fig. 6. Simulation of the pathway of the fabrics for concept 2

The first trials showed problems, that the scrims could not be fed into the die due to their relatively high built-up in dry condition. To compensate for this, the scrim layup was changed so that only one scrim per guide slot was used. A fiber content of 50% was aimed for. The pultrusion of these profiles worked, but the FVC was too low, resulting in poor surface quality (Fig. 9 right).

There were also problems, such as the shifting of the scrims in the guide slots due to the too large height, which led to wrinkles. In an iterative process, the guiding panels were redesigned and optimized with the help of a simulated guiding of the scrims (Fig. 6). To ensure a good transition and preforming, the guide panels were geometrically illustrated by CAD with the expected course of the scrims, also to ensure that there are no collisions between the layers.

In the first trial (concept 1), nearly all guiding panels (see Fig. 4, green) were made of polyoxymethylene (POM) and the core bearing panels were made of steel. To have a smoother transition of the straight fabrics from the resin bath to the closed loop of the tube surface section, the pathway for concept 2 was drawn nearer the core to have smaller preforming form panel to panel. As a result of this redesign, in which the scrims are led closer along the core, the new panels were made of aluminum instead of POM to reduce the risk of deformation due to high process forces.

To have enough space for the uncompacted scrims and to ensure as few wrinkling as possible, additional additive manufactured elastic rings were designed and manufactured as shown in Fig. 7. As an option, it is possible to leave these out to allow more space around the core to lead in the scrims. The tests showed good results of the usage of those rings.

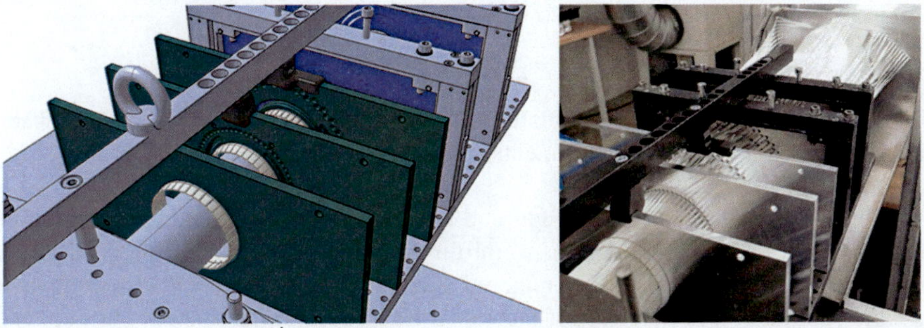

Fig. 7. Guiding panels with elastic rings for concept 2; CAD (left); Experiment (right)

3 Materials and Processes

In the following chapter, the focus is on the manufacturing process for the fiber-reinforced tubes.

3.1 Process

In the pultrusion process, profiles with complex cross section can be produced very economically. The standard process for manufacturing straight profiles with thermoset matrix is shown in Fig. 8: Semi-finished fiber products (1) are pulled from bobbins by alternately moving pulling devices (4) and pass through a resin bath (2). Afterwards, the impregnated fibers are pulled through a heated die (3), in which the liquid thermoset plastic cures completely within seconds to a solid plastic. A saw (5) cuts the profiles to the desired length [15].

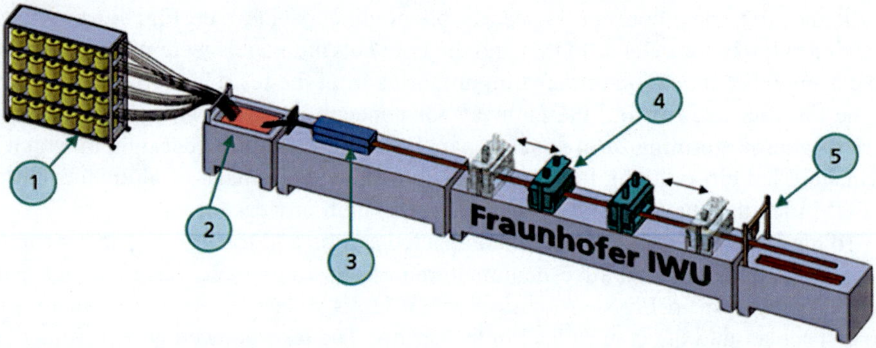

Fig. 8. Scheme of the pultrusion line

3.2 Materials

To achieve the designed and calculated failure modes and the needed energy absorption for the lightweight crash elements, the crash tubes are supposed to be made of CFRP.

However, for the feasibility study and technology development made on the pultrusion line at the Fraunhofer IWU, semi-finished products made of glass fibers were used. Due to the small amount of material needed for the experiments (in comparison to series production) a purchase of fabrics with the desired (simulated) built up, especially with the fiber angles of $\pm 30°$ in CRFP wasn't available in the timeline of the project. To evaluate a feasibility of pultrusion of these crash tubes, glass fiber fabrics are equally suitable—a transfer of the experience gained to CFRP will take place in a post project to evaluate the crash properties in comparison to the simulation.

The semifinished fiber material used is a triaxial scrim from the manufacturer SAERTEX with the fiber directions 0° (378 g/m^2), +30° (366 g/m^2) and −30° (366 g/m^2). Rovings used for the outer and inner layer were of the type PulStrand 4100 with 4,800 tex from the manufacturer Owens Corning. As matrix system an unsaturated polyester was used. Table 1 shows the components of the polymer system.

Table 1. Resin system

Component	Component name	Supplier
Resin 1 Resin 2	Synthopan 781–60 Synthopan 134–61	Synthopol Synthopol
Curing agent 1 Curing Agent 2	Peroxan MI-60 KX Peroxan BEC	Pergan Pergan
Monostyrene	Monostyrene	BÜFA
Internal Mold Release	PAT-667	Würtz
Inhibitor	Pergaslow PK-30S	Pergan

4 Results

The overview of the main steps for the experimental setup is stated in Table 2.

Table 2. Overview experimental setup

No	Guiding panels	Layup	FVC	Result
1	Set 1 (Concept 1)	Inside: 29 Rovings Middle: 8 fabrics Outside: 65 Rovings	60%	Not possible to feed the fabrics
2	Set 1 (Concept 1)	Inside: 23 Rovings Middle: 4 fabrics Outside: 66 Rovings	50%	Pultruded tubes with poor surface due to low FVC (Fig. 9), wrinkling, dry spots
3	Set 2 (Concept 2)	Inside: 60 Rovings Middle: 5 fabrics Outside: 65 Rovings	55%	Pultruded tubes non-concentric, less wrinkling
4	Set 2 (Concept 2)	Inside: 30 Rovings Middle: 6 fabrics Outside: 44 Rovings	58%	Pultruded tubes (Fig. 10)

Figure 9 shows the results of the first trials (Table 2, No. 2). Due to difficulties in guiding the scrims and fibers, a FVC of about 50% could be achieved. Because of this low FVC, there was a very poor profile quality.

Fig. 9. Setup (left) and result (right) of trial no. 2

As described in Sect. 2.2, several optimization steps were carried out to achieve a much better result. The pultrusion line was built up with the new guiding system and at first tested with a layup of rovings, 5 layers of fabric and rovings. The results were positive in general, a tube with a FVC of about 55% could be pultruded (Table 2, No. 3) – but due to the relatively large number of rovings, there was some difficulty in achieving a concentric tube. In the next trial (Table 2, No. 4), the layup with 6 layers of fabric and less rovings on the inside and outside was successfully tested. The result is shown in Fig. 10.

Fig. 10. Setup (left) and result (middle and right) of trial no. 4

5 Discussion and Conclusion

As part of a modular vehicle structure, it was shown that a manufacturing of crash tubes via pultrusion as a series-suitable process is possible with various adaptions of the process. Various concepts were created for the crash structure on the modular design side, one of which was selected and refined using CAD modeling and FEM simulations. For the manufacturing of the tubes, two concepts for the layer-built up were discussed more closely—closed layers or layers consisting of two halves of fabric. The chosen one was with closed loops, to have less seams for a better performance and less influence of the fabric's borders. Afterwards, the pultrusion die and guiding panels were designed and produced. The first production implementation showed several problems, like feeding the scrims into the die, wrinkling and a poor surface quality due to a too low FVC. The findings were then implemented in a new design of guiding planes and additional elastic support rings.

The next step after this project is to transform the know-how of pultruding a tube with a complex layup by using carbon fibers and carbon scrims. When transferring the technology developed here, it may be necessary to consider using an injection box for impregnation. This would facilitate processability for the significantly more sensitive carbon semifinished products and fibers.

Acknowledgements. The project "KOSEL" is funded within the funding measure "Resource-efficient Circular Economy—Innovative Product Cycles (ReziProK)". "ReziProK" is part of the research concept „Resource-efficient Circular Economy" of the Federal Ministry of Education and Research (BMBF) and supports projects that develop business models, design concepts or digital technologies for closed product cycles.

SPONSORED BY THE

Federal Ministry
of Education
and Research

References

1. Finnveden, G., et al.: Recent developments in life cycle assessment. J. Environ. Manage. **91**(1), 1–21 (2009)
2. Hellweg, S., Milà i Canals, L.: Emerging approaches, challenges and opportunities in life cycle assessment. Science **344**(6188), 1109–1113 (2014)
3. Holzhauer, R.: Altauto-Demontage: Bisherige Entwicklungen und Realität. Recycling und Rohstoffe; In: Thomé-Kozmiensky, K.J., Goldmann, D. (eds) pp. 151–171 (2015)
4. Fraunhofer IWU, EDAG Engineering GmbH: KOSEL – Reusable open source design kit for electrically powered pool vehicles. https://innovative-produktkreislaeufe.de/resswinn/en/Projects/KOSEL.html. Accessed 14 Feb 2022
5. Crespo, S. et. al.: Structural Optimisation in Vehicle Development for the current Euro NCAP side crash protocol: how to minimise the structural changes due to the current barrier stiffness and geometry. In: Proceedings of IRCOBI Conference-Malaga, number IRC-16-82 (2016)
6. Van Paepegem, W. et al.: Blast performance of a sacrificial cladding with composite tubes for protection of civil engineering structures. Compos. B. Eng. **65**, 131–146 (2014)
7. Hull, D.: A unified approach to progressive crushing of fibre-reinforced composite tubes. Compos. Sci. Technol. **40**(4), 377–421 (1991)
8. Charoenphan, S., Bank, Lawrence C., Plesha, M. E.: Progressive tearing failure in pultruded composite material tubes. Composite Structures **63**(1), 45–52 (2004)

9. Nossol, P., Saleem, A.: Experimentelle Untersuchung zu faserverbundgerechten Crash-Wirkmechanismen bei axialer Druckbelastung. Konstruktion 3, 7582. issn: (Print): 0373-3300, (Electronic): 0720-5953 (2020)
10. Kakogiannis, D., Chung Kim Yuen, S., Palanivelu, S., van Hemelrijck, D., van Paepegem, W., Wastiels, J., Vantomme, J. u. Nurick, G. N.: Response of pultruded composite tubes subjected to dynamic and impulsive axial loading. Compos. B. Eng. **55**, 537–547 (2013)
11. Ding, L., Liu, X., Wang, X., Huang, H. u., Wu, Z.: Mechanical properties of pultruded basalt fiber-reinforced polymer tube under axial tension and compression. Constr Build Mater. **176**, 629–637 (2018)
12. Nossol, P., Drebenstedt, C., Szymanski, F.: Design and manufacture of long-lasting lightweight structures for sustainable mobility. Technologies for Lightweight Structures (TLS) **5**(1), 15–23 (2022)
13. Nossol, P.: Smarte Crash-Wirkprinzipien für hybride Fahrzeugstrukturen. SmartPro Workshop, Chemnitz (2018)
14. Euro NCAP: Oblique pole side impact, https://cdn.euroncap.com/media/53145/euro-ncap-pole-test-protocol-v71.pdf. Accessed 18 Feb 2022
15. Starr, T.F.: Pultrusion for engineers. Cambridge, Boca Raton, Fla.: Woodhead; CRC Press (2000)

Thermocouple Fabrication by Cold Plasma Spray

N. Mainusch[1,2(✉)], D. Scholz[1,2], J. Linkmann[2], T. Abraham[1,3], and W. Viöl[1,2]

[1] Fraunhofer Institute for Surface Engineering and Thin Films IST, Von-Ossietzky-Street 99-100, 37085 Göttingen, Germany
{nils.mainusch,daniel.scholz, wolfgang.vioel}@ist.fraunhofer.de, tim.abraham@tu-braunschweig.de

[2] HAWK Hochschule für angewandte Wissenschaft und Kunst, Und Kunst Von-Ossietzky-Street 100, 37085 Göttingen, Germany
jakob.linkmann1@hawk.de,

[3] Technische Universität Braunschweig, Langer Kamp 19B, 38106 Braunschweig, Germany

Abstract. Fraunhofer IST and University of Applied Sciences and Arts (HAWK) have recently developed a Cold Plasma Spray (CPS) coating technique providing for superior assembly of microsensors and other delicate substrates or devices. In CPS systems pulsed voltages at low currents ignite transient arc discharges that result in thermally moderate plasmas. Further, CPS allows for the use of air or nitrogen as plasma gas and provides a considerably gentle coating deposition process enabling the metallization of heat-sensitive or mechanical-weak objects. This is in contrast to conventional High-Temperature Thermal Plasma Spray or High Velocity Oxygen Fuel Spraying. The CPS process extends to a myriad of delicate surface applications, including the fabrication of electrically conductive bars and thermal sensors such as accurate thermocouples. These elements can be deposited onto thin polymer foils, e.g., ≤10 μm-thick PET, glass optics, or organic coatings; the so-called "cathodic dip paint" which is common in the automotive industry. Moreover, depending on the substrate, the CPS process allows for device extensions permitting robot-based metallization without the need for masking. The hereinafter presented work elucidates the fabrication of thin-film thermocouples by means of the unique CPS plasma spray process at Fraunhofer IST with emphasis on two use-cases.

Keywords: Plasma Spray · Sensor · Thermocouple · Metallization · Copper · Constantan

K. Dröder and T. Vietor (Eds.): *Future Automotive Production Conference 2022*, Zukunftstechnologien für den multifunktionalen Leichtbau, pp. 59–70, 2023.
https://doi.org/10.1007/978-3-658-39928-3_5

1 Introduction

Many technical applications are encumbered by heat fluxes that affect the performance of the final electrical/optical components and systems. For example, irradiated light emitting lenses or protective glasses are sensitive to thermal stresses. Such conditions result in undesired property changes. Arising from thermally induced size increases their refractive indexes can be detrimentally modified. At worst, these influences degrade and even completely destroy the elements. In order to circumvent heat-induced impairment of technical glasses, optics, or electrical components as well as to define their operating limits property characterizations require reliable material temperature measurements. Besides this, the need for a precise *in-situ* temperature monitoring can be exemplified by electric vehicle batteries and compensatory temperature measurements for sensors such as strain gauges. Challenges that go along with accurate temperature recording and the state-of-the art of corresponding devices will be clarified in the following.

Different physical phenomena are utilized for temperature measurements. Examples include the temperature-driven electrical resistance of conductors, and the generation of thermoelectric force with respect to specific metals during joining operations. The current state-of-the-art devices are resistance thermometers and thermocouples. Thermocouples are comparative measurement methods. They are comprised of two dissimilar metallic wires which form an electrical junction. This results in a quasilinear temperature-dependent voltage. The typical physical realization of this device is a tip offering locally resolved absolute temperature values measured against a reference junction. Thermocouples benefit from robustness against environmental impacts. Indeed, this provides for uncomplicated implementation even over long distances between the measuring tip and the data interface. Various types of thermocouples operate over wide temperature ranges (e.g., type K: from −270 °C to 1300 °C, type T: from −270 °C to 400 °C). Of course, a great advantage of thermocouples is economical: they are inexpensive. Unfortunately, while thermocouples offer a precise and economic sensor technology, their accuracy under *in operando* conditions is reliant upon careful positioning and sensor-to-substrate contact. In many cases, a high temporal resolution is an additional challenge. The current state-of-the-art suffers from inefficiencies; when the simple intimate contacting of an object becomes intricate. That is, not only must the junction of the thermocouples and standard wires/probes be welded together, but the device must be mounted and intimately attached to the substrate of interest (Fig. 1).

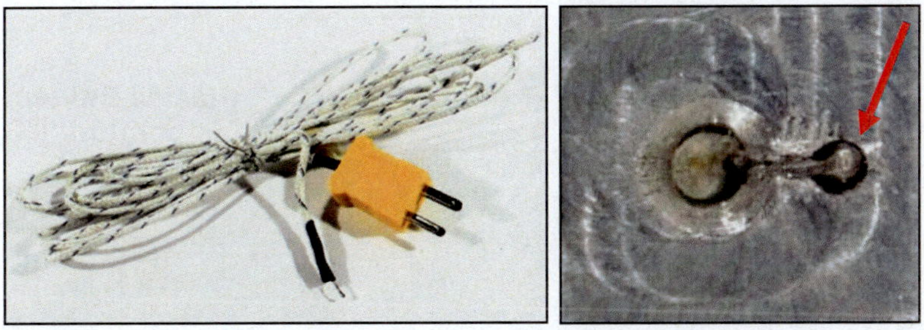

Fig. 1. (LEFT) a commercially available welded-tip Type K thermocouple, and (RIGHT) inserted (arrow) into a holder for in operando thermal measurements. This set-up aims to provide intimate contact of the obverse of a silicon wafer mounted into a pocket.

Thin-film thermocouples provide a pathway to overcome these bottlenecks. The current thin-film manufacturing techniques include: Calendaring, Physical Vapor Deposition (PVD), and Thermal Plasma Spray [1]. The first can be exemplified by introducing a thermocouple with flat wires between, e.g., polyimide sheets and attach this composite onto an object [2, 3]. Drawbacks include the requirements of polymers and adhesives with restricted thermal stabilities. PVD deposited thermocouples on customized substrates are commercially available [4]. Here, limitations exist because substrates might be inadequate for vacuum treatment, vacuum generation is expensive, time-consuming, and masking requirements. A more advanced method for manufacturing thin film thermocouples is Thermal Plasma Spray. [5–7]. Nevertheless, this technology suffers from the high thermal loading of substrates during deposition, and the time-consuming requirement of masking. Existing shortcomings of conventional (thin-film) TC fabrication/implementation include problems due to inappropriate positioning and bad sensor-to-substrate attachment, contact loss during operation, high TC production costs, and risk of substrate damage in fabrication. In comparison, the Cold Plasma Spray technique can circumvent these bottlenecks by reducing, or outright eliminating, these traditional processing requirements.

2 Cold Plasma Spray

Cold Plasma Spray (CPS) is a thermo-kinetic coating technology operating at atmospheric pressure with fine powders as precursors. Unlike Thermal Plasma Spray or High Velocity Oxygen Fuel Spray, CPS features a thermally moderate plasma effluent and low particle velocities. A comparison of the characteristic temperatures and particle velocities of the methods is given in Fig. 2.

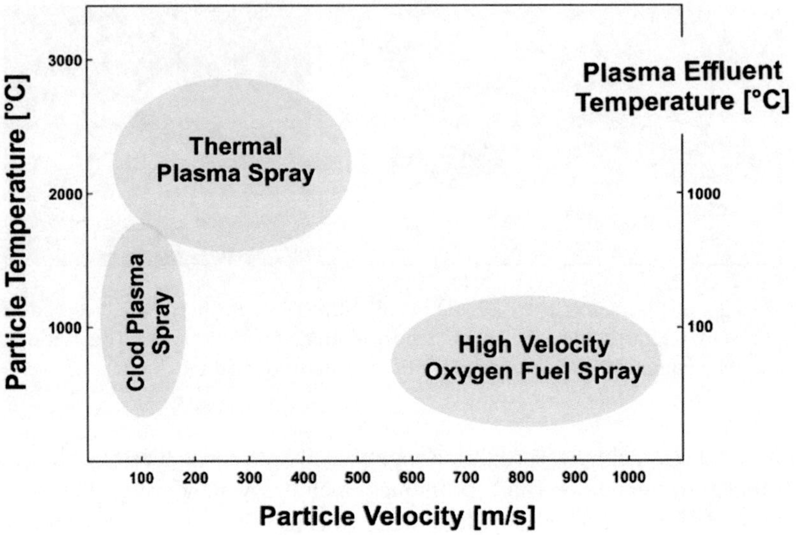

Fig. 2. Comparison of typical particle temperature ranges, velocities and plasma effluent temperatures for Cold Plasma Spray (own measurements) against similar thermo-kinetic coating technologies (personally communicated by experts from different R + D institutes).

The CPS generates plasma jets by means of pulsed DC-discharge voltages of 3–4 kV at frequencies of approx. 50 kHz. The electrical input power is less than 1 kW; significantly lower than Thermal Plasma Spray (50–100 kW). The low input power results in comparatively moderate effluent temperatures, minimization of thermal impact damage of substrates, thereby allowing for the metallization of thermally-delicate polymer foils. A major advantage of CPS at Fraunhofer IST/HAWK (System: Plasma Plotter 3D, INOCON Technologies GmbH, Austria) is that this technology allows for the implementation of much finer powders than conventional spray systems (mean particle diameter, $d_{50} \ll 5$ µm; compared with typically ≈ 10 µm) and specific particle morphologies (platelet-like). These precursors cannot be processed by standard powder feeders. At IST/HAWK advanced feeders were developed. These facilitate steady aerosol generation with a wide range of ultra-fine powders (virtually any metal and many organic powders, as well). Due to size and shape CPS powders possess a high specific surface area. This promotes superior thermalization of the precursor and permits to reduce plasma energy input and accordingly, low jet temperatures. Nitrogen, forming gas, or air serve as both the ionization gas and the particle carrier medium. This is an advantage over Thermal Plasma Spray which requires noble ionization gases (typically argon or helium). A simplified schematic of the CPS nozzle and associated process is illustrated in Fig. 3.

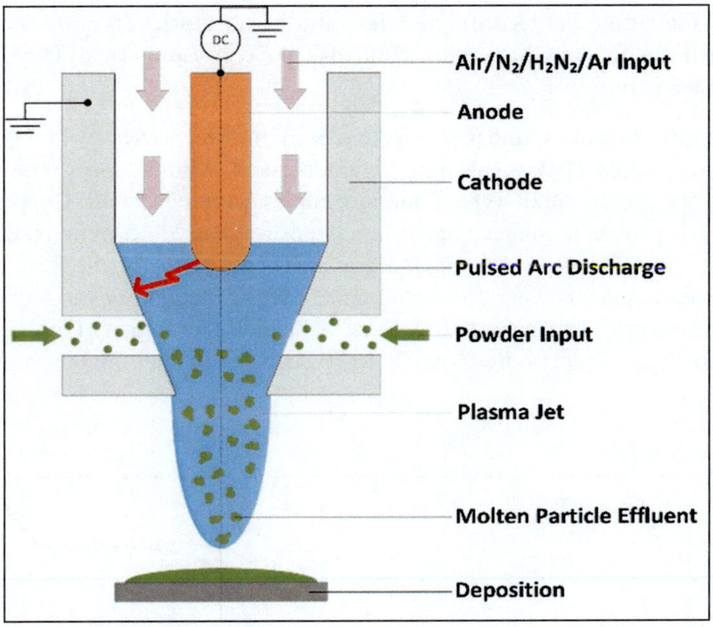

Fig. 3. CPS nozzle and process.

3 Use-Cases, Results and Discussion

Preliminary investigations were conducted using a wide array of powder materials and substrates to explore both their processability and applicability for a wide gamut of applications. This included CPS of copper, bismuth, and constantan tracks. Test substrates were thin polymer foils (PET, PE, etc., 50–100 µm thickness) and painted metal plates. The qualities of these plasma-metallicized surfaces were evaluated according to both track adhesion and electronic conductivities. Further, the thermoelectrical potential of resulting metal pairs were checked. To provide a realistic measure of commercial/industrial applicability, the dielectric stability of the heat-sensitive polyurethane paint, thermo-pair (Cu-Bi, Cu-Cu/Ni) functionalities, and the integrities of the thin foils were interrogated. The summation of these tests demonstrated that CPS-deposition of thermocouples onto delicate substrates is feasible.

In an effort to further promote the process regarding industrial demands and use-cases, Cu-Cu/Ni thermocouple fabrications onto glasses and on a cathodic dip coating were conducted. Within comparative measurements thin-film *versus* commercials sensor responses were examined. The effect of convection and irradiation as heat sources, and process upgradability in terms of a new approach for a "selective metallization without masking" were explored. Details of these studies are presented below.

3.1 Thermocouple Fabrication on Glass and Laser Optics Dynamics of a Thin-Film TC on Glass and a Welded-Tip TC in Long-Term Heating by Convection

Thermocouple dynamics under convective heat transfer were evaluated *via* long-term measurements. Glass substrates were masked with heat-resistant polyimide tape and copper/constantan Type T thermocouples were deposited. Copper and constantan wires provided connectivity to the deposited bars. A commercially available welded-tip Type T class 1 sensor served as the benchmark for the CPS devices. In all cases, measurements were conducted under identical conditions in a drying chamber. Measurements were recorded with an 8-channel data logger (Pico USB TC-08) while ramping the temperatures from 20 to 90 °C at 0.5 °C per minute as presented in Fig. 4.

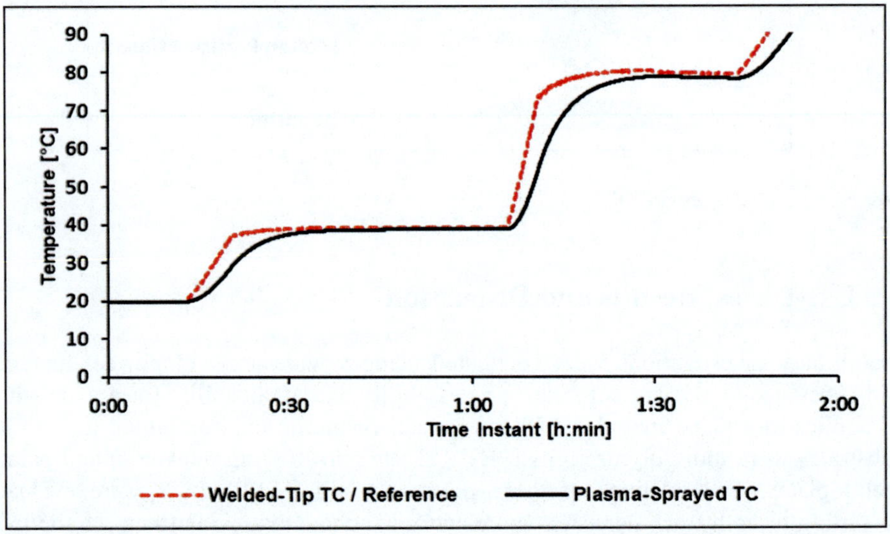

Fig. 4. Long-term heating by convection comparing a thin-film plasma-sprayed TC on a glass sphere (solid line) and a conventional welded-tip TC (dashed line). The latter was not in contact with the glass, but separately situated.

The curves represent temporal responses for the plasma-sprayed TC in comparison with the conventional welded-tip sensor. That is, the plasma-sprayed TC exhibits a comparatively smooth measured temperature increase compared to the welded-tip TC (dashed line). Maintaining constant heat input effectuates that after some 10 min. All sensed temperatures are matched. Apparently, the plasma-sprayed TC reacts slower than the welded-tip TC. This might be explained by the higher mass of the glass-thermocouple-composite in comparison with the finite welded-tip TC which requires less heat quantity in order to grow warm. Generally, in this set-up the tip TC turns out to be advantageous in comparison with the thin-film TC. To further elucidate the response of the plasma-sprayed TC additional tests with short-term laser heating were carried out.

Dynamics of a plasma-sprayed thin-film TC and a welded-tip TC being pressed on optics in rapid heating by laser irradiation

Akin to the experiment described above, a copper and a constantan bar (width ~ 1 mm) were cross-wise plasma-sprayed onto an optical lens (diameter ~ 14.6 mm, thickness ~ 2.5 mm). Thin wires of likewise copper and constantan were compacted onto one flank of each bar and connected with the data logger. In contrast to depositing and wiring, the welded-tip TC was pressed towards the optics' surface and situated as close as possible to the thin-film sensor (Fig. 5) and connected by the existing connector cable.

Fig. 5. Glass sphere fitted into a holder (white cylinder). The thin-film TC is located on the backside of the mounted glass. Two looped wires (arrows) made of copper and constantan contact the bars. One welded-tip TC (dashed arrow) is pressed against glass' backside close to the intersection of the deposited bars.

The thermal loading was provided by CO_2 laser irradiation (TruCoax). An approx. 14 mm laser beam (Gaussian profile) with inclined incidence was focused onto the front side of the glass providing 200 ms of irradiation at a power of 120 W. Each pulse was followed by a cooling-off time of 60 s. Multiple irradiation and relaxation cycles were conducted to explore the effect of the optics. These results are summarized in Figs. 6 and 7.

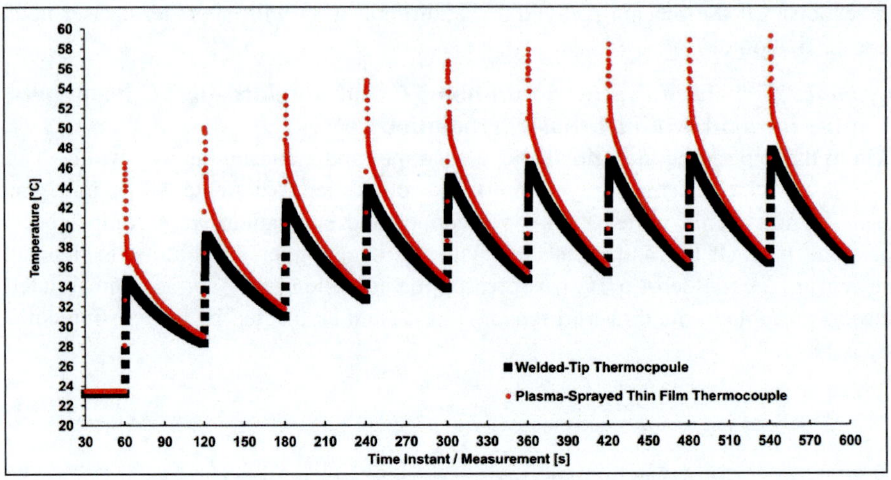

Fig. 6. Measurement results comparing a thin-film TC (red dotted line) and a welded-tip TC (black squares), both sensors being positioned according to Fig. 5. Object heating by laser irradiation was 200 ms and cooling-off time was 60 s. A total of 9 cycles were conducted.

Figure 6 indicates that over the course of the cycles a thermal saturation and almost equilibrium for both TCs is achieved. However, while the thin-film TC approaches approx. 60 °C, the welded-tip TC exhibited an increase of approx. Only 48 °C. Thus, a minor absolute value for the welded-tip TC can be declared.

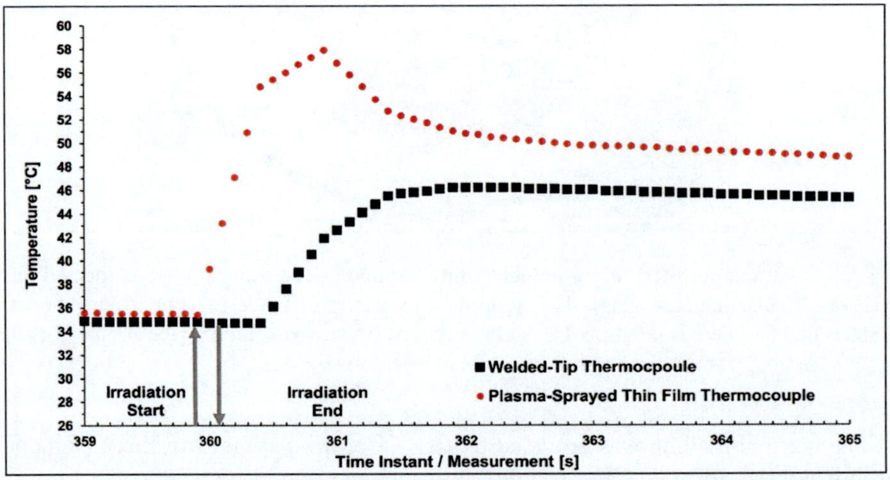

Fig. 7. Exemplary measurement result according to Fig. 6 (6th cycle) between time instants 359 and 365 s with 0.1 s time-resolution, thin-film TC (dotted line) and welded-tip TC (squares). The irradiation starting point is indicated with an ascending arrow at the time instant 359.9 s. The base temperature between the sensors differ in approx. 1 °C due to previous heating and cooling-off cycles.

Moreover, Fig. 7 clearly indicates that the laser irradiation boosts the thin-film TC response immediately to approx. 56 °C after the laser excitation. The sensed temperature increases a while lineally after the cessation of the pulse and by time instant approx. 360.6 s continues to rise, but at a lower slope. The temperature thereafter drops after approx. 1 s, and settles down after roughly 2 s towards a value of 50 °C. In contrast, the welded-tip TC does not respond until after the end of the irradiation period. Moreover, the temperature increase is considerably muted; barely any declination can be observed. The curve finally approaches equilibrium at approx. 45 °C. The welded-tip TC is clearly less dynamic over short timeframes than the thin-film TC. The reliable detection of rapid thermal changes is crucial with respect to mechanical stresses and resultant formation of cracks. Indeed, the structural continuity of glasses is severely tested by high heating/cooling velocities. CPS-deposited TCs provide the resolution to monitor these thermal fluxes owing to their intimate contact with the substrate. The explanation for quicker CPS TC response time is its good contact to the object.

3.2 Selective Metallization/Thermocouples on Cathodic-Dip-Paint

In the automotive industry, so-called cathodic-dip-paint (CDP) is a well-established process for the anti-corrosion protection of truck chassis and analogous applications. CDP polymer coatings typically feature a thickness of 20 µm. These coated chassis are subjected to mechanical stresses when trucks drive and its chassis move. Chassis system control has therefore traditionally required the implementation of resistance strain gauges. Further, accurate thermal measurements are needed to both decrease calibration times and increase precision of resistance strain gauges. Thermocouples are cost-effective appropriate sensors for these applications.

It was previously suggested that one objective of this use-case involved the realization of selective metallization without the need to mask the substrate. A promising pathway would be to upgrade the plasma-spray process by implementing a micro-sandblasting device. The intention is to locally roughen the surface to generate a surface that promotes adhesion of the deposition material. Adverse to this, the non-blasted, pristine CDP remains smooth. By continuously controlling the plasma-spray parameters it was hypothesized that poor adhesion of metal deposition occurs in unroughened zones. This would allow for realizing metal tracks along only previously structured tracks.

A Resko Airblaster II sandblaster was outfitted with a finite nozzle (diameter 0.8 mm) which was aligned with the plasma-spray jet. Consecutively, single pass sandblasting (spray medium was corundum with grid size 240) and metallization were executed via a robot-controlled process. The arrangement is depicted in Fig. 8.

Fig. 8. Metal substrate coated with a cathodic dip paint (black) mounted on a robot-controlled sample stage. Selective metallization is accomplished by; 1) meticulously sandblasting and roughening the paint (thin nozzle in front), and 2) depositing copper and constantan via plasma deposition. Note the adhesion promoting structures (grey lines) with a width of approx. 1 mm and the as-deposited copper thermo legs (brown). Lateral track accuracy is within 0.2 mm.

Figure 9 presents a typical cross-section offering insight into the composition of the copper thermo leg. Owing to the sandblasted surface, the copper layer is partially embedded into the CDP promoting adhesion. The coating thickness of the 20 µm CDP was relatively unaffected by the process, and the dielectric insulation of the metal plate was preserved. Comparative testing of the manufactured thin-film TC and a conventional welded-tip TC attached to the CDP under convective heating (oven) resulted in identical measuring values and equal accuracies (validated by measurements, result not shown here). Ergo selective roughening of CDP by micro-sandblasting provides a simple and effective method for the fabrication of metal tracks and the generation of TCs without masking. Further minimization of the nozzle (diameter <0.8 mm) will permit both smaller tracks $<$ approx. 1 mm while increasing the accuracy of the lateral tracks as desired.

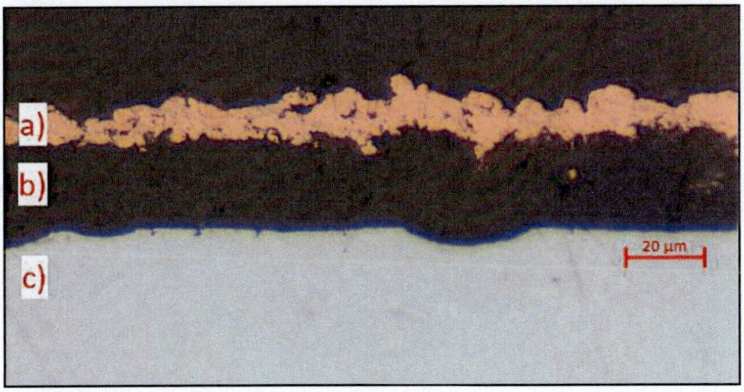

Fig. 9. Cross section of a plasma-sprayed copper thermo leg. a) copper, b) CDP, c) steel plate.

4 Conclusions

These preliminary studies demonstrate that Cold Plasma Spray for fabricating copper and constantan tracks, as well as type T thermocouples, on heat-sensitive coatings, polymer foils, and other delicate substrates is feasible. While conventional welded-tip TCs provide sufficiently accurate temperature measurements, the CPS-deposited devices offer much higher temporal resolutions required for delicate operations. In general, CPS-deposited TCs offer improved reliability with respect to the absolute temperature value of an object and higher accuracy under pulsed thermal loads. An immediate application for this technology is in the realm of qualifying laser optics or similar optical elements. Finally, a simple but yet efficient method for the precise and selective metallization of cathodic dip paint by means of micro sandblasting in conjunction with CPS was demonstrated. This supersedes the masking and facilitates three-dimensional objects to be equipped with electric tracks and sensors, and is an advantage compared to the current-state-of-the-art.

Acknoledgements. The authors thank the German Research Foundation (Deutsche Forschungsgemeinschaft, DFG) for funding the plasma spray system (INST 196/11–1 FUGG).

References

1. Liu et al.: A thin-film temperature sensor based on a flexible electrode and substrate. Microsyst. Nanoeng. **7**, 42 (2021)
2. https://www.tcdirect.net.au/Default.aspx?level=2&department_id=180/47#180/47/1 visited: March 2022. https://www.tcdirect.net.au/Default.aspx?level=2&department_id=180/47#180/47/1
3. Fluxteq, https://de.fluxteq.com/ftc-foil-thermocouples, TCdirect, visited: March 2022.
4. Kyodo-inc: https://www.kyodo-inc.co.jp/english/electronics/wafer-process/data/thermistor.html, visited: March 2022

5. Gutleber, J. et al.: Embedded temperature and heat flux sensors for advanced health monitoring of turbine engine components. 2006 IEEE Aerospace Conference, p. 9 (2006) https://doi.org/10.1109/AERO.2006.1656117
6. Mitchel, D.,et al.: Development and F-Class Industrial Gas Turbine Engine Testing … . Proceedings of ASME Turbo Expo 2008: Power for Land, Sea and Air GT2008 June 9–13, Berlin (2008)
7. Longtin, J., et al.: Sensors for Harsh Environments by Direct-Write Thermal Spray. IEEE SENSORS Journal, **4**(1) (2004)

Life Cycle Engineering

A Variability Model for Individual Life Cycle Paths in Life Cycle Engineering

Lukas Block[1,1(✉)], Maximilian Werner[2], Helge Spindler[1], and Benjamin Schneider[1]

[1] Fraunhofer Institute for Industrial Engineering (IAO), Nobelstraße 12, 70569, Stuttgart, Germany
{lukas.block,helge.spindler, benjamin.schneider}@iao.fraunhofer.de

[2] Institute of Human Factors and Technology Management (IAT), University of Stuttgart, Nobelstraße 12, 70569, Stuttgart, Germany
maximilian-jakob.werner@iat.uni-stuttgart.de

Abstract. Life cycle properties are becoming increasingly important for the success of a product. They are determined during development, but their impact—and thus value—only becomes apparent in later life cycle phases. Life cycle engineering is concerned with designing such product properties in early stages of design. However, the life cycle paths of individual products from the same product type increasingly diverge, due to differing product usage, personalization and software or hardware updates. The expected and realized value of certain life cycle properties might vary greatly within the same type of product. Thus, this paper addresses how such individual life cycle paths can be made accessible to product system designers in the context of LCE, to evaluate and determine valuable life cycle properties. A meta model is developed to describe divergent life cycle paths of individual products and investigate, how to identify the possible, future life cycle paths in early stages of design and how to incorporate them in LCE. The approach is applied to the design of a door panel for passenger cars. Suitable life cycle properties and their associated designs were evaluated and determined, with respect to the expected, individual life cycle paths.

Keywords: Life Cycle Properties · Meta model · Product development · Model-based Systems Engineering

1 Introduction

The evolution of a product during its life cycle is central to life cycle engineering (LCE). LCE is an engineering approach, which seeks to incorporate life cycle properties into a product in early stages of design [1]. Life cycle properties are non-functional properties of a product system or its parts. They are determined during development.

K. Dröder and T. Vietor (Eds.): *Future Automotive Production Conference 2022, Zukunftstechnologien für den multifunktionalen Leichtbau*, pp. 73–85, 2023.
https://doi.org/10.1007/978-3-658-39928-3_6

Yet, their impact—and thus value—only becomes apparent in later life cycle phases [1, 2]. Examples are customizability, serviceability, reusability as well as the product's recyclability (see e.g. [1–3]). Life cycle properties are becoming increasingly important for the success of a product [4, 5]. On the one hand, additional life cycle properties such as reusability, repairability, and remanufacturability currently come to the fore due to the increasing demand for sustainable and circular products [6, 7]. On the other hand, product costing focus shifts from a development and manufacturing perspective to an overall life cycle perspective due to Servitization and the development of product-service-systems (see e.g. [8]). Life cycle engineering focuses on determining such properties in the early (concept) stage of product development, when product attributes and costs can still be influenced to a large extend [1]. Thereby, rebound effects as well as further dependencies between the life cycle properties and other design goals must be considered [3, 9]. Current methods and tools for LCE support in identifying and resolving such trade-offs during product development. They analyze the life cycle of the product under development and identify valuable designs for the life cycle properties.

However, the life cycles of individual products from the same product type increasingly diverge. A natural source of this divergence are the circumstances of the individual product usage. They define, how long an individual product lasts, whether it is maintained, resold, put to a second-life use, or recycled. Additionally, mechatronic product systems like cars are increasingly personalized and updated. They evolve individually during the life cycle through reconfiguration, software updates and hardware upgrades. This further increases product variations in the field (see e.g. [10]). As a result, the expected and realized value of certain life cycle properties might vary greatly within the same type of product. A design problem in the conceptual phase of product development results, regarding LCE: A system concept must be determined, which effectively leverages the life cycle properties over all possible, individual life cycle paths. To achieve this, the individual paths must already be known and considered in concept development. Otherwise, the value of certain life cycle properties cannot be evaluated properly. Rebound effects, occurring between individual life cycle paths, are ignored and the life cycle properties are not realized effectively. Trade-offs between the properties and their suitability for different life cycle paths are not identified and resolved. Thus, we address how such individual life cycle paths can be made accessible to product system designers in the context of LCE. Firstly, a meta model is developed to describe divergent life cycle paths of individual products. Subsequently, we investigate, how to identify the possible, future life cycle paths in early stages of design and how to incorporate them in LCE.

2 Literature Review

Jeswiet [11] provides a fundamental definition of the term Life Cycle Engineering (LCE), summarizing forgoing research: Life Cycle Engineering sums the "engineering activities which include the application of technological and scientific principles to manufacturing products with the goal of protecting the environment, conserving resources, encouraging economic progress, keeping in mind social concerns, and the need for sustainability, while optimizing the product life cycle and minimizing pollution and waste.". Thus, product development plays a central role in LCE as this phase defines e.g. a product's environmental footprint in large part [12]. Consequently, LCE can be understood to comprise product development activities and decisions, keeping certain product properties such as sustainability aspects in sight while considering single or multiple product life cycles [13–16]. Pivotal elements of LCE are life cycle properties. "These properties are not the primary functional requirements of a system's performance, but typically concern wider system impacts with respect to time and stakeholders that are embodied in those primary functional requirements" [2]. Typical life cycle properties, which de Weck et al. [2] describe as "-ilities", are amongst others "Reliability", "Flexibility", "Durability" and "Usability". Regarding product systems that are supposed to fulfill requirements of a Circular Economy, these life cycle properties could be enhanced or derived, by considering circular strategies like "Repurpose", "Refurbish", "Re-use" and "Rethink" (see e.g., [6]).

LCE has to face new challenges in terms of considering additional life cycle phases and states in product design, for instance resulting from sustainability-oriented concepts such as Circular Economy [7, 17, 18]. This is caused by the need to extend product life cycles to keep products, components, and materials in use as long as possible. This is expedient with the requirement to consider additional phases of e.g. product usage already in product development [19, 20]. Thus, the Circular Economy paradigm amongst others represents a challenge for engineers to integrate a holistic life cycle thinking approach [9]. This is exemplified by the research of Halstenberg et al. [8], who developed a methodology for the development of Smart Services, which addresses Circular Economy strategies. They propose Model-based Systems Engineering (MBSE) procedures, notations and tools as an adequate foundation [8]. Model-based Systems Engineering (MBSE) describes the model-based and IT-supported application of systems engineering methods [21] to optimize modeling and foster a common understanding and a traceability for the system under development [22, 23]. As such, holistic approaches for MBSE generally consist of a method, a (software) tool and a (graphical) modeling language [24], which are in this paper's focus. Other methods, pointing in the same direction as Halstenberg et al. [8] with MBSE for LCE, are for example Bougain et al. [25] and Yvars et al. [26]. Thereby, Cerdas et al. [27], Dér et al. [28], and Tao et al. [29] for example take divergent, individual life cycle circumstances in LCE into account. Yet, their perspective is data focused and simulation driven. They are not suitable for early stages of design, when this data is not present and detailed simulations are not expedient. Thus, our analysis led us to the insight that there is a need for product development, and LCE in particular, to take the alteration of individual life cycles over time into account. To the authors' understanding, this has not been covered by the findings of Halstenberg et al. [8], Cerdas et al. [27], Dér et al. [28], and Tao et al. [29].

To summarize, different approaches for the evaluation of life cycle properties as well as methodologies for life cycle-oriented design exist. They analyze and model the general product's life cycle during development or evaluate the individual life cycles in detail via simulations and collected data. Yet, to the best of our knowledge, none of them proactively incorporates divergent life cycle paths of individual products in early stages of concept development to determine life cycle properties. As such, we aim to answer the following research question: How can divergent and individual life cycle paths be described and incorporated in LCE, to evaluate and determine valuable life cycle properties?

3 Approach

The goal of our approach is to incorporate individual life cycle paths into LCE. The life cycle paths are employed to determine and evaluate valuable life cycle properties with respect to all possible, individual life cycle paths. The literature review indicates that there is a sufficiently large and profound body of knowledge, dealing with LCE approaches based on generic (i.e., non-individual) life cycles. They build up on a systemic and model-based understanding of the product and its life cycle. Thus, we follow an orthogonal approach. The already existing methodologies and models are supplemented by developing a meta model for individual life cycle paths. This meta model can be used as a complement to already existing product and life cycle models. Subsequently, the meta model is associated with a methodology, which describes, how to setup and utilize such a model in present LCE approaches.

Overall, our approach is based on the idea of MBSE. The individual life cycle paths are described as models. This creates transparency by abstracting the most important aspects of the individual life cycle paths. Furthermore, it allows analysis of the modelled paths and interference towards the influence on the life cycle properties. Systems engineering provides the conceptual framework to link our approach with the already present LCE approaches [9, 30].

3.1 Meta Model for Individual Life Cycle Paths

In MBSE, a model stores all the available information and knowledge about the product to be developed [22, 23]. The systems engineering methodology then generates, interacts, and alters the model throughout the design process, to integrate information and knowledge and evaluate the current design. A meta model defines the representation scheme for the information and knowledge within the product model. For LCE such a representation scheme should include, among other things, information and knowledge about the life cycle performance of the product (see e.g., [27–29]). Thereby, one challenge is, that the associated, individual life cycle paths must be depicted in early stages of design, before the product is developed. The individual life cycle paths are only partially known and uncertain.

We define an individual life cycle path as a sequence of multiple life cycle states. A life cycle state is a temporal demarcated, feasible situation, in which specific properties of the product are requested by external stakeholders. These external stakeholders are for example the customer, the legislative body, or the manufacturing department. The union of all required properties from all life cycle states equals the overall product requirements. A life cycle phase aggregates the life cycle states, whose required properties are driven by the same set of external stakeholders. Thus, multiple life cycle states for one life cycle phase can coexist.

In contrast, a life cycle property is a non-functional, system wide property, whose impact only becomes apparent in later life cycle phases [1, 2] (see Sect. 1 and 2). Life cycle properties are realized through specialized approaches (see e.g. [8, 25, 26]), which translate the abstract objective of the life cycle property into a product's design. Thereby it is important to design the product in such a way, that it generates a real positive impact on later life cycle phases. Yet, the life cycle phases consist of different life cycle states, which might be passed individually by each product. Thus, incorporating a life cycle property in the product's design is only of value, if it supports the realization of the required properties of a sufficiently large portion of the life cycle states. Put vice versa, the required properties of the life cycle states define the value of a life cycle property and how it is best implemented. Thus, the life cycle properties must be evaluated and implemented according to the given life cycle states and their probabilities. This is supported, by modeling the life cycle states' probabilities as well as the relationship between life cycle states and life cycle properties explicitly.

Figure 1 depicts the elements of our meta model. Conceptually, a model for LCE with individual life cycle paths is divided into three parts: The product architecture, the variability model and the life cycle model. Each part describes a certain aspect of possible product states throughout the life cycle phases. The product architecture models the requirements, functions, components and generic physical structure of the product to be developed (see [10, 31]). It represents the static part of the design. We assume the meta model for the product architecture is externally given and thus not part of our meta model. Yet, the variability and life cycle model of our meta model reference elements of the architecture model, which are expected to differ between the life cycle states. The variability model defines the required properties which might change individually during the vehicle's lifetime or throughout the life cycle states. The required properties are depicted either as requirements or as functions, components, and modules, which are necessary to fulfill the states' required properties. Thereby, elements of the product architecture model, which might change, are referenced by var-

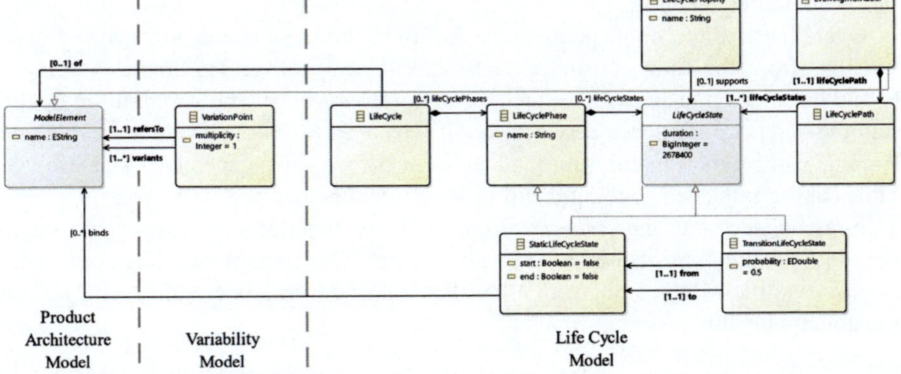

Fig. 1. The most important elements and attributes of the meta model with its three parts.

iation points. Possible realizations of these elements are depicted as variants of these variation points. Interdependencies between the variants are modelled according to product line engineering principles. The variability model is an optional part of our meta model because some architecture meta models already support variants.

In general, the variability model does not assign the variants to any specific life cycle state nor to any individual product. This is realized via the life cycle model. The life cycle model references the variants in the variability model and bundles them to consistent, static life cycle states. Static life cycle states describe a certain and fixed-in-time configuration of requirements, functions, and components, which are specific for this state. Yet, transitions between the static life cycle states might occur, which represent a dynamic state of product evolution (see Fig. 3). According to the definition of a life cycle state, such a transition is also a life cycle state. Thus, a transition life cycle state is additionally introduced, which describes the product's state while evolving from a previous static state to a next static state. Transition probabilities for the transition life cycle states allow to model the dependencies and underlaying logic, regarding which life cycle state might later be realized for the individuals. Thereby, transition probabilities are understood as in the Bayesian interpretation of probability. The transition probabilities are prior estimates of the proportion of products produced, which will take this transition (see [32]). LCE incorporates life cycle properties in early stages of design. As such, the prior estimates must build upon subjective beliefs or previous, yet not immediately transferrable knowledge. The prior estimates might thus stem for example from predecessor products, market studies, simulations, or personal experience of the product designers. They can even be expressed through more rough and uncertain probability measures like fuzzy set theory, possibility theory or Dempster Shafer evidence theory (see e.g., [33]). Probability distributions can then be derived from them.

Finally, life cycle properties are modelled qualitatively by the objective they aim to achieve. They are associated with life cycle states, in which they might provide a value by supporting or realizing the required properties therein. The association is then used in the methodology to determine and evaluate valuable life cycle properties (see Sect. 3.2). The life cycle states, and their transitions form a Markov chain (see Fig. 3). Exemplary life cycles for individual products can be derived through Monte-Carlo-Simulation (see Fig. 4).

Overall, the three conceptual parts decouple the design and life cycle property decisions from the time and individual-focused perspective. This reduces complexity and enhances transparency as well as comprehensibility. The variability model for example depicts differences between the life cycle states in the product architecture's domain, while ignoring the time and individual-focused perspective. The life cycle states can be modelled without a full product architecture available. Yet, they can be easily associated with the respective requirements, functions or even components of a product architecture, if they are already present. Our meta model has been defined according to the MOF standard to allow for maximum interoperability with different product architecture models.

3.2 Methodology and Model Perspectives

The meta model's conceptual structure already helps in reducing complexity when dealing with individual life cycle paths. Yet, a methodology is necessary. It should

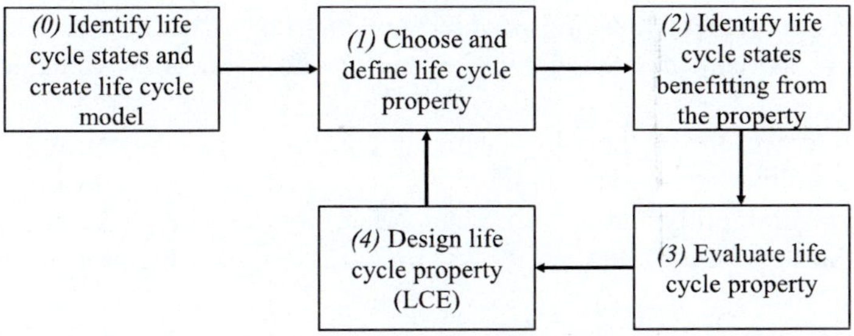

Fig. 2. Methodology to incorporate individual and divergent life cycle paths into LCE.

describe how to setup a life cycle model from our meta model and how to use it to evaluate and determine the life cycle properties. Our proposed methodology is depicted in Fig. 2.

Initially, the life cycle model must be setup. This is already a small subordinate methodology on its own. In a first step, the life cycle states are defined. For each life cycle phase, the product's stakeholders in this phase as well as their requested product properties are identified. The product properties are then aggregated, based on the situation in which they become relevant. These situations are the static life cycle states. Transitions between the life cycle states are determined, based on hypothetically feasible product evolutions. The transition probabilities are derived from previous products, market research or experience for example (see Sect. 3.1). Finally, the life cycle model should be checked. The probability distribution over all life cycle states can be controlled to lay within the expected ranges, and individual life cycle paths can be investigated regarding their feasibility (see Figs. 4 and 7).

The life cycle model can then be used, to evaluate and determine the life cycle properties. Firstly, potential, beneficial life cycle properties for the product are identified. They can be derived for example from customer demands and market trends or the company's strategy. Subsequently, each potential life cycle property is associated with a set of life cycle states. These are the states, in which the life cycle property might deliver value. As such, each required property of each life cycle state is compared against the potential life cycle property. If the states' required properties are supported by the life cycle property, the state is associated with the life cycle property (see Fig. 4). Favorable technical implementations of the life cycle property become apparent.

Thirdly, the life cycle properties are evaluated. The probabilities that a life cycle state benefitting from the life cycle property is passed in an average, individual life cycle is calculated. This proxy for the value of a life cycle property helps in deciding to which extend the life cycle property should be incorporated. High probabilities hint towards a higher utilization of the designed properties. Low probabilities suggest that the life cycle property is only of value for a small portion of the individual products. Ultimately, the life cycle property can be designed into the product architecture using a suitable LCE methodology from literature. Thereby, knowledge regarding a beneficial design is already present from step (2). Furthermore, exemplary, individual life cycles and the probability distribution of the life cycle states can be used, to evaluate design solutions in this process. Consequently, the approach extends LCE: The meta model supplements existing product models to account for diverging life cycle paths and an associated methodology evaluates and determines valuable life cycle properties.

We supplement the methodology with three prototypical viewpoints of our models to reduce cognitive complexity. Perspective (1) depicts the potential life cycle states and their transition probabilities as product variability over time (see Fig. 3). It depicts different states and links them via the required properties to the product architecture.

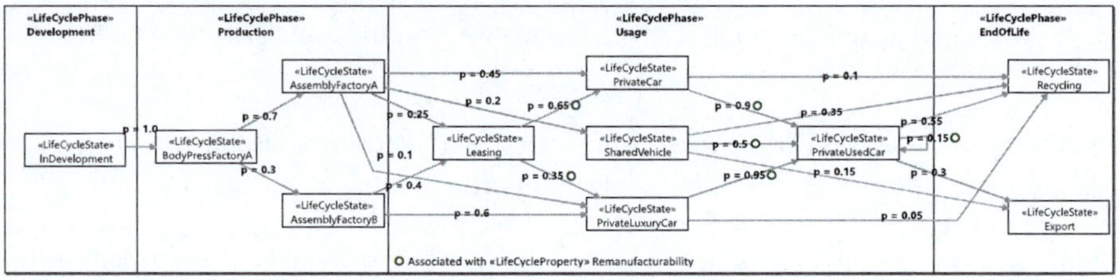

Fig. 3. Perspective (1) depicts the life cycle states and the transition probabilities between them (exemplary data from the evaluation).

Thus, it helps to uncover possible variants in the product architecture and is suitable to design the life cycle model. This perspective is mainly employed in the initial step of the methodology as well as in step (2).

The remaining perspectives (2) and (3) focus on the product to be developed. Perspective (2) depicts an exemplary life cycle path for a certain individual product

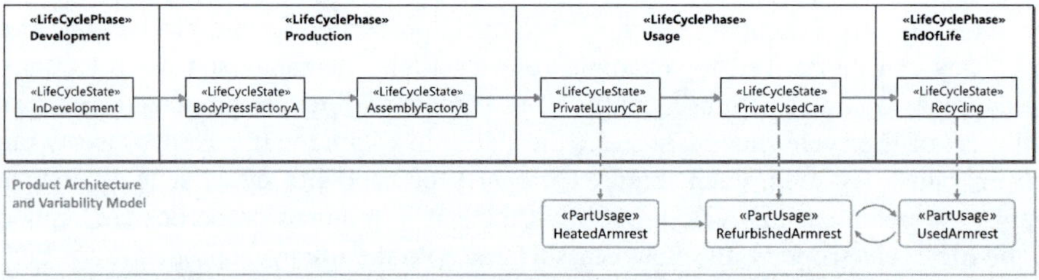

Fig. 4. Perspective (2) depicts exemplary life cycles (i.e., a specific, probable sequence of life cycle states) for individual products (exemplary data from the evaluation).

over time (see Fig. 4). The probabilistic interdependencies are abstracted to infer implications towards the design decisions in step (4) of the methodology. Thereby, it is important to generate sufficient diverse and representative life cycles.

Perspective (3) is arranged around the product architecture. It depicts the product's requirements, functions and components with their different variants (see Fig. 5). They stem from the properties, required in the individual life cycle states. This

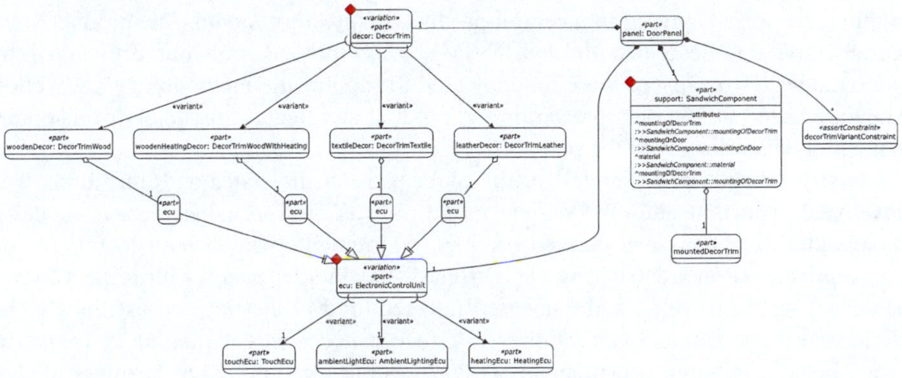

Fig. 5. Perspective (3) depicts the product architecture with its life cycle state induced variability (exemplary data from the evaluation); thereby, variability is marked with a red diamond.

perspective reveals points of potential change throughout the life cycle phases but ignores the interdependencies and sequence of life cycle states. It is used in step (4) of the methodology.

4 Evaluation and Discussion of Case Study Results

The evaluation of the meta model and the associated methodology has been conducted in the research project "futureFlexPro" [34]. The research project aimed for developing a flexible and sustainable smart door panel (SDP) for passenger cars. The meta model was implemented into an Eclipse modelling environment to employ the developed meta model and approach. The modelling environment features SysML v2 for the product architecture description. Overall, the software architecture of the modelling environment looked as follows (see Fig. 6): The information and knowledge about the required properties, life cycle states, life cycle properties and the product

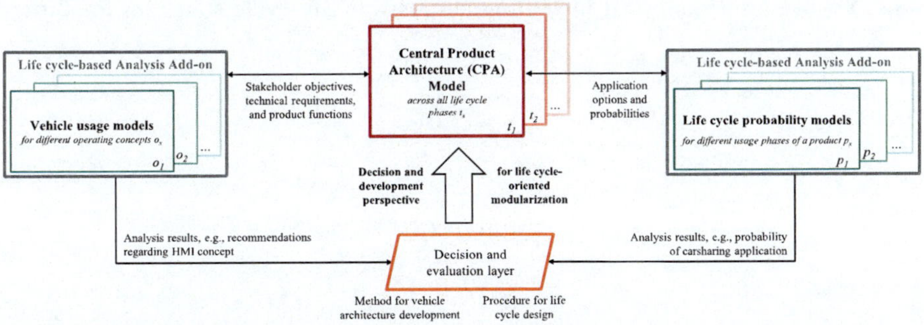

Fig. 6. The evaluation's software architecture for modelling and evaluating a smart door panel.

architecture were captured in a central product architecture model. The product architecture model is described through SysML v2 combined with our orthogonal life cycle model. Two add-ons were implemented to support the methodology's workflow. They are based on the data present in the product architecture model. A decision and evaluation layer displayed the perspectives, presented in Sect. 3.

Firstly, the life cycle model for the passenger car, its states and transitions were developed. The first add-on "Vehicle usage models" supported in this. A model of stakeholder objectives and derived product requirements was developed for two use cases: private car and carsharing car. Different stakeholder needs with respect to both use cases were identified and converted into required properties and assigned to life cycle states. For instance, an intuitive interface concept was requested by carsharing users, because reading user manuals is not feasible for time-based business models. Private car users in contrast were linked to shy tech and individualizable configurations. They can precisely adapt to their specific needs. As a car is either part of a shared fleet or used as a private car, different life cycle states were defined including the required properties of the stakeholders (see Fig. 3). Subsequently, the transition probabilities were derived based on different experts' experience. At the same time, the life cycle model was associated with the product architecture. The need to exchange parts of interface components, for example when the vehicle is owned by a private user after being used for carsharing, became apparent.

The probability distributions for each life cycle state and each life cycle path were derived based on the life cycle model through the add-on "Life cycle probability models" (see Fig. 7). Potential beneficial life cycle properties regarding for example usability, reliability, performance, changeability, and remanufacturability were identified. They were associated with the life cycle states (see e.g., Fig. 4) and evaluated. It was found for example, that different SDP configurations throughout the life cycle of the same vehicle were required with a high probability (i.e., 89%). This in turn, pointed the designer towards the value of the life cycle properties remanufacturability and changeability for the SDP. They were realized for example for the décor material using magnetic coupling instead of simple adhesive joints. Thereby, perspective (3) helped to identify the points of change. Other recommendations derived concerned the placement of the electronic control units or necessary material characteristics of the SDP. Overall, our approach made it possible to analyze frequency, probability, and type of potential component exchanges, based on individual life cycle path estimations. Yet, we also found that modelling all possible life cycle states for the different

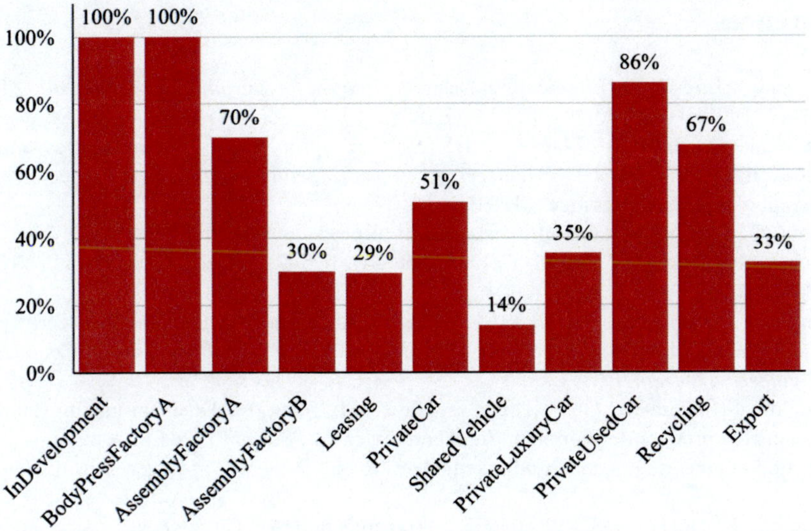

Fig. 7. Probabilities of the life cycle states, occurring at least once in an individual life cycle path.

stakeholders and linking them to the life cycle properties is cumbersome. Thus, our upcoming research regarding this topic, will address this issue, by extracting knowledge from previous linkages and automate the tool support.

5 Conclusions

The goal of this paper is to add the consideration of individual life cycle paths to LCE, so that valuable life cycle properties for divergent life cycle paths can be evaluated and determined. We suggested a new meta model to describe individual life cycles in LCE. The meta model is complemented with a methodology. It describes how to use the meta model to evaluate and determine valuable life cycle properties and their implementation. Thereby, the approach supplements already existing LCE approaches. The meta model and the methodology can be used for different life cycle properties, given that a MSBE focused LCE approach for them already exists. In an exemplary evaluation case, our approach was applied to the design of a sustainable smart door panel. Suitable life cycle properties and their associated designs were evaluated and determined, with respect to the expected, individual life cycle paths.

Acknowledgements. The authors gratefully thank the German Federal Ministry of Education and Research (BMBF) for funding (funding code: L1FHG42421).

References

1. K. Ishii, "Life-Cycle Engineering Design," *Journal of Vibration and Acoustics*, 117, B, 42–47 (1995). https://doi.org/10.1115/1.2838675
2. de Weck, O., Roos, D., Magee, C. L.: Life-Cycle properties of engineering systems: The Ilities. In O. de Weck (ed.) *Engineering systems: Meeting human needs in a complex technological world.* Cambridge: MIT Press (2011)
3. Feng, C., Mai, Y.: Sustainability assessment of products based on fuzzy multi-criteria decision analysis. J. Adv. Manuf. Technol. **85**(1–4), 695–710 (2015). https://doi.org/10.1007/s00170-015-7978-1
4. de Weck, O., Eckert, C., Clarkson, P.: A classification of uncertainty for early product and system design. In *Proceedings of ICED 2007: The 16th International Conference on Engineering Design*. Paris (2007)
5. Lindow, K., Heimann, O., Adolphy, S., Hayka, H., Stark, R.: Decision-making support for sustainable product development. In: Abramovici, M., Stark, R. (eds.) Lecture notes in production engineering, smart product engineering, pp. 979–988. Springer, Berlin, Heidelberg (2013)
6. Potting, J., Hekkert, M., Worrell, E., Hanemaaijer, A.: *Circular Economy: Measuring Innovation in the Product Chain: Policy Report* (2017)
7. Aguiar, M.F., Mesa, J.A., Jugend, D., Pinheiro, M.A.P., Fiorini, P.D.C.: Circular product design: strategies, challenges and relationships with new product development," *MEQ*, ahead-of-print, ahead-of-print (2021) https://doi.org/10.1108/MEQ-06-2021-0125
8. Halstenberg, F.A., Lindow, K., Stark, R.: Leveraging circular economy through a methodology for smart service systems engineering. Sustainability **11**(13), 3517 (2019). https://doi.org/10.3390/su11133517
9. Herrmann, C.: Ganzheitliches life cycle management. Springer, Berlin, Heidelberg (2010)
10. Block, L.: Guiding local design decisions towards a flexible and changeable product architecture. In *Proceedings of the Design Society: DESIGN Conference*, pp. 521–530 (2020)
11. Jeswiet, J.: Life cycle engineering. In: Laperrière, L., Reinhart, G., Chatti, S., Tolio, T. (eds.) Springer reference, CIRP encyclopedia of production engineering, pp. 757–758. Springer, Berlin (2014)
12. Hauschild, M.Z., Herrmann, C., Kara, S.: An integrated framework for life cycle engineering. Procedia CIRP **61**, 2–9 (2017). https://doi.org/10.1016/j.procir.2016.11.257
13. Hauschild, M.Z., Rosenbaum, R.K., Olsen, S.I. (eds.): Life cycle assessment: Theory and practice. Springer, Cham (2018)
14. Hauschild, M.Z., Kara, S., Røpke, I.: Absolute sustainability: Challenges to life cycle engineering. CIRP Ann. **69**(2), 533–553 (2020). https://doi.org/10.1016/j.cirp.2020.05.004
15. Wanyama, W., Ertas, A., Zhang, H.-C., Ekwaro-Osire, S.: Life-cycle engineering: Issues, tools and research. Int. J. Comput. Integr. Manuf. **16**(4–5), 307–316 (2003). https://doi.org/10.1080/0951192031000089255
16. Laurent, A., et al.: The role of life cycle engineering (LCE) in meeting the sustainable development goals: Report from a consultation of LCE experts. J. Clean. Prod. **230**, 378–382 (2019). https://doi.org/10.1016/j.jclepro.2019.05.129
17. de Schoenmakere, M., Gillabel, J.: *Circular by design: Products in the circular economy*. Luxembourg: Publications Office of the European Union (2017). http://publications.europa.eu/en/publication-detail/-/publication/84527ea6-861a-11e7-b5c6-01aa75ed71a1

18. Mestre, A., Cooper, T.: Circular product design. A multiple loops life cycle design approach for the circular economy. Des. J. **20**(sup1), S1620–S1635 (2017). https://doi.org/10.1080/14606925.2017.1352686
19. den Hollander, M.C., Bakker, C.A., Hultink, E.J.: Product design in a circular economy: Development of a typology of key concepts and terms. J. Ind. Ecol. **21**(3), 517–525 (2017). https://doi.org/10.1111/jiec.12610
20. Diaz, A., Schöggl, J.-P., Reyes, T., Baumgartner, R.J.: Sustainable product development in a circular economy: Implications for products, actors, decision-making support and lifecycle information management. Sustainable Production and Consumption **26**, 1031–1045 (2021). https://doi.org/10.1016/j.spc.2020.12.044
21. Sillitto, H., Martin, J., McKinney, D., Griego, R., Dori, D., Krob, D.: *Systems Engineering and System Definitions..* https://www.incose.org/docs/default-source/default-document-library/incose-se-definitions-tp-2020-002-06.pdf. Accessed 11 Feb 2022
22. Dumitrescu, R., Albers, A., Riedel, O., Stark, R., Gausemeier, J.: *Engineering in Deutschland - Status quo in Wirtschaft und Wissenschaft: Ein Beitrag zum Advanced Systems Engineering*. Paderborn (2021)
23. INCOSE - International Council on Systems Engineering: SE Vision (2025) https://www.incose.org/products-and-publications/se-vision-2025
24. Friedenthal, S., Moore, A., Steiner, R.: A practical guide to SysML: The systems modeling language. Morgan Kaufmann Publishers Inc, San Francisco (2014)
25. Bougain, S., Gerhard, D.: Integrating environmental impacts with SysML in MBSE methods. Procedia CIRP **61**, 715–720 (2017). https://doi.org/10.1016/j.procir.2016.11.196
26. Yvars, P.-A., Zimmer, L.: A model-based synthesis approach to system design correct by construction under environmental impact requirements. Procedia CIRP **103**, 85–90 (2021). https://doi.org/10.1016/j.procir.2021.10.013
27. Cerdas, F., Thiede, S., Herrmann, C.: Integrated computational life cycle engineering: Application to the case of electric vehicles. CIRP Ann. **67**(1), 25–28 (2018). https://doi.org/10.1016/j.cirp.2018.04.052
28. Dér, A. Kaluza, A. Reimer, L. Herrmann, C., Thiede, S.: Integration of energy oriented manufacturing simulation into the life cycle evaluation of lightweight body parts. *Int. J. of Precis. Eng. and Manuf.-Green Tech.* (2022). https://doi.org/10.1007/s40684-021-00412-w.
29. Tao, J., Yu, S.: A meta-model based approach for LCA-oriented product data management. Procedia CIRP **69**, 423–428 (2018). https://doi.org/10.1016/j.procir.2017.12.010
30. Gelfert, A.: *How to do science with models*. Cham: Springer International Publishing (2016)
31. Ulrich, K.: The role of product architecture in the manufacturing firm. Res. Policy **24**(3), 419–440 (1995). https://doi.org/10.1016/0048-7333(94)00775-3
32. Beck, J.L.: Bayesian system identification based on probability logic. Struct. Control Health Monit. **17**(7), 825–847 (2010). https://doi.org/10.1002/stc.424
33. Ayyub, B.M., Klir, G.J.: Uncertainty modeling and analysis in engineering and the sciences. Chapman & Hall/CRC, Boca Raton (2006)
34. Hipke, T.: *futureFlexPro: Entwicklung variantenflexibler und ökoeffizienter Systemkomponenten für kommende Fahrzeuggenerationen im Sinne einer ganzheitlichen Kreislaufwirtschaft..* https://www.hybridleichtbau.fraunhofer.de/de/forschungsprojekte/futureflexpro.html. Accessed 7 Feb 2022

Increase the Ressource Efficiency by Evaluation of the Effects of Deep Rolling within the Design and Manufacturing Phase

Oliver Maiß[1,1(✉)], Karsten Röttger[1], and Kolja Meyer[2]

[1] ECOROLL AG Werkzeugtechnik, Hans-Henrich-Warnke-Street 8, 29227, Celle, Germany
{oliver.maiss,karsten.roettger}@ecoroll.de

[2] Technische Universität Braunschweig, Technische Universität Braunschweig, 38106, Braunschweig, Germany
kolja.meyer@tu-braunschweig.de

Abstract. Car manufacturers are currently facing the same challenge as all manufacturing companies: To increase the resource efficiency associated with their products. However, the demands are somewhat higher for car manufacturers: They must reduce greenhouse gas emission during their cars' use phase. But it is not enough if only the use phase is adapted, but as a manufacturer, the OEMs must also include the emissions during production into their considerations. The consequence of this will be that the supply chain has to increase the resource efficiency. Resource efficiency can be increased by lightweight design of components. An idea that has been implemented for many years is lightweight construction by using the design of surface integrity. By optimizing surfaces and near surface regions, the service life of dynamically loaded components can be increased by introducing compressive residual stresses. Processes used for this are processes such as deep rolling or machine hammer peening. However, the effects of these processes directly on the reduction of part weight and the resulting effect on the production chain have not been investigated sufficiently. Within this paper, the possibilities of lightweight construction by deep rolling will be investigated based on an experiment and on literature data. For this purpose, dynamic loading tests are carried out on three different component states and thus a weight saving with the same service life is enabled. Subsequently, an evaluation of the material weight reduction is performed. These investigations will be the base for further investigation regarding a possible CO_2 emission reduction along the production chain.

Keywords: Deep rolling · Lightweight design · Life cycle analysis · CO_2 emissions

K. Dröder and T. Vietor (Eds.): *Future Automotive Production Conference 2022, Zukunftstechnologien für den multifunktionalen Leichtbau*, pp. 86–96, 2023.
https://doi.org/10.1007/978-3-658-39928-3_7

1 Introduction

Manufacturing of passenger vehicles is an important industrial sector in Germany. In the year 2016, 30% of the global revenue for car sales was created by German car manufacturers [1]. In the year 2012, 47% of German electrical energy consumption was used for industrial applications, from which 23.9% were used for metal production and 12.9% were used for construction of machines and motor vehicles [2]. One method for the reduction of energy usage and therefore CO_2 emissions lies in lightweight construction of the used parts [3]. Lightweight construction can be achieved by the usage of mechanically stronger materials, which also results in higher cost for the material itself. Another option is the design of the parts surface integrity. Surface integrity is defined as the surface area and surface near subsurface region which are influenced by machining processes [4, 5]. Using adapted operations and parameters, the surface integrity state can benefit the static and dynamic strength of loaded components. Especially during cyclical loading, the service life is strongly influenced by the surface integrity state. For bending rotating loads, the fatigue strength is mostly influenced by the hardness, phase composition, residual stress state, and roughness [6]. In general, a lower roughness with a high valley radius and high distance between as well as compressive residual stresses are seen as beneficial for the service behavior of cyclically loaded parts [7]. With high compressive residual stresses, the roughness influence is reduced [8, 9]. For another important loading this effect was also shown. For roller bearings, Voskamp identified the phases of residual stress states: Increasing compressive residual stresses, steady-state phase and uncontrolled changes leading to damage [10, 11]. Hacke showed that inducing more compressive stresses by a preloading, the lifetime of bearings is significant higher [12]. The surface integrity state is influenced by the mechanical, thermal, and chemical loads during machining [13], therefore, finishing operations play a main role in surface integrity design as they are the last machining step before the use stage. Currently, it is still challenging to predict the surface integrity state, because for most machining processes, an overlap between the different load categories exists [14]. One mostly mechanical finishing process is deep rolling or burnishing. In the process a roller is pressed with a rolling force onto a surface by using hydraulic or mechanical systems. Within the contact area of the roller and the surface very high contact stresses occur and the surface and subsurface area is plastically deformed, which leads to reduced surface roughness, compressive residual stresses and an increased hardness. During deep rolling, only minimal thermal loads are applied to the workpiece, it can therefore be described by the mechanical contact between workpiece and tool [15]. Because of the existing process knowledge, deep rolling processes can be used to set the residual stresses [9, 16] and the roughness [9] to a beneficial state which increases service life during mechanical loading. Meyer shows that with the introduction of compressive residual stresses using deep rolling, the service life of rotating bending loaded workpieces can increase by $\Delta\sigma_B = 178\%$ [9]. For roller bearings, deep rolling is able to induce exactly the necessary amount of residual stresses for a service life enhancement. The values and the depth of the residual stress can be induced where it is needed to get the steady-state phase. Within the combination with the actual bearing loads the compressive residual stresses decrease the von-Mises stresses which leads to an increase of the service life by about + 150% [16–18].

While an increase in service life can be seen as beneficial, another benefit of the surface design induced increase of material resistance can be the option to reduce the part size and weight. The application of tailored surface integrity states can therefore be used to enable lightweight applications without the necessity for higher grade materials. In this paper, this concept is applied to bending rotating loaded parts and to bearings.

2 Concept to Evaluate the Resource Efficiency for Deep Rolled Parts

As described in the introduction, deep rolling induces compressive residual stresses which are able to increase the fatigue life of dynamically loaded parts. This fact is used as a motivation for lightweight in several scientific projects for many years. However, the important question is, how this improvement can be used in reality for increasing the resource efficiency and reducing CO_2 emissions within the manufacturing and operation phase of a part. To answer this question, this paper describes a practical method to evaluate the amount of resources and CO_2 emissions which can be saved due to deep rolling. Therefore, first the method will be presented and demonstrated with two different parts and loadings afterwards.

To use the potentials of deep rolling to reduce CO_2 emissions, two different approaches can be applied. The first one is to simply use the parts for a longer time and use fewer spare parts. If a parts fatigue life is improved by the factor of 2 it can be used twice as long. Which is great for parts considered as spare parts. That means we need to produce half the number of parts and half the amount of material is used. However, this is not applied very often, because the customer is not willing to pay the higher price. To gain the same amount of profit with half the produced parts the manufacturer has to increase prices. And there is another issue with this approach. Most parts are designed to be used as long-life parts, which are not supposed to be damaged during the lifetime of a machine at all.

Therefore, the second approach seems to be more beneficial for reducing overall CO_2 emissions. If the lifetime of a part is longer, the part can easily be designed smaller and reaches the initial fatigue life. If the part is smaller, it will use less material, less weight must be carried around and the surrounding parts can also be smaller.

The approach of this paper is demonstrated in Fig. 1. First the part, the kind of loading and the necessary fatigue life are defined. Then one important step is to identify the optimal process parameters for the deep rolling process. To identify the optimum, costs, productivity and properties have to be considered. Here it is possible to find parameters in experiments or use the knowledge of experts and literature sources.

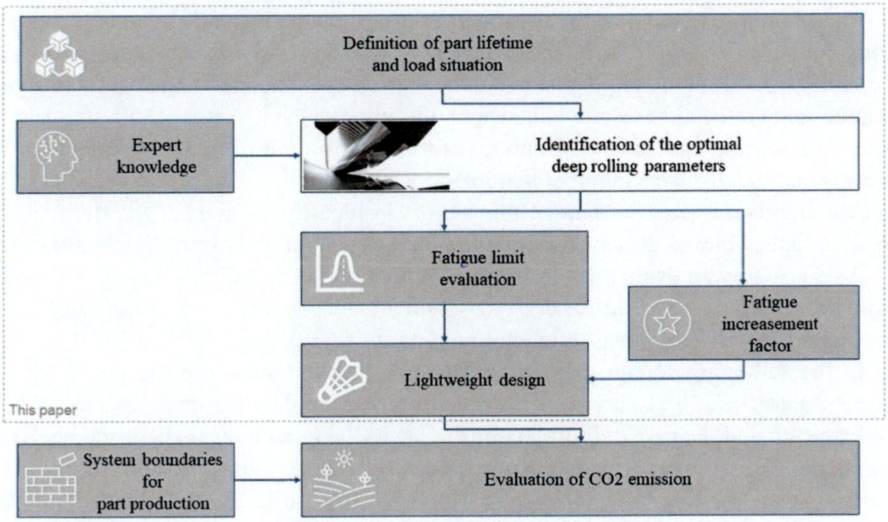

Fig. 1. Concept for the evaluation of the CO_2 reduction due to lightweight design.

After the machining parameters are identified, fatigue tests are carried out. Here the improvement factor for the specific part must be found. Therefore, a comparison between a conventional manufactured and deep rolled part needs to be drawn. As shown in Fig. 2, these two parts have the same weight but one part last longer with a specific fatigue life improvement.

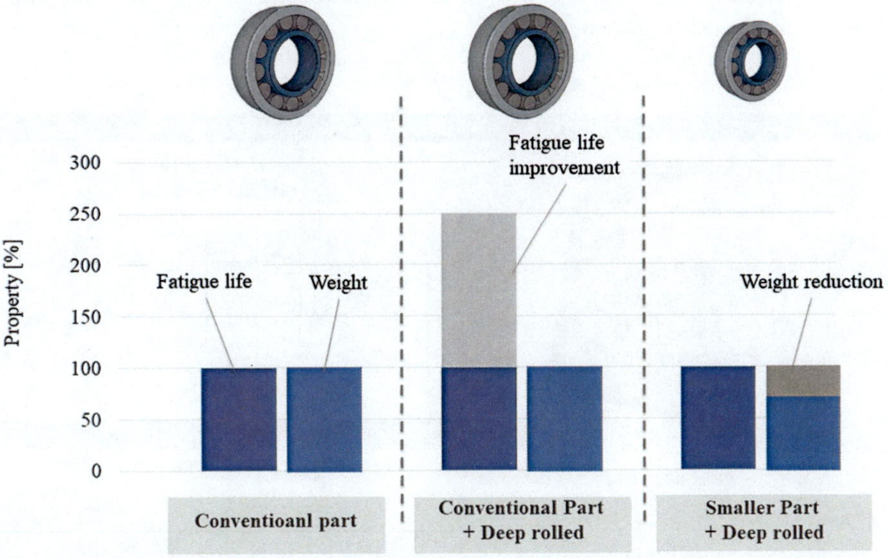

Fig. 2. Lightweight design by deep rolling.

The life improvement can be identified in experiments or can be assumed from previous parts. For example, the factors can be stored in databases in the future. Using this component, a lightweight design can be carried out. Therefore, the part is designed smaller, and the lifetime is calculated with the improvement factor. Again, this can be done very simple by using conventional models and multiply this lifetime with the improvement factor. Or elaborated simulations and fatigue tests can be carried out, to find the lighter design. The aim of this step is to identify the part of the right side of Fig. 2, which achieves the same fatigue life but has a significant weight reduction.

As a last step an evaluation of the CO_2 reduction is carried out, which can be performed by application of a life cycle assessment (LCA). This step is not included in this paper but will be addressed in later publications.

In the following sections of this paper, the different steps are discussed for two different parts. The first part is a roller bearing. All the data for this are coming from the literature and the research project of the Deutsche Forschungsgemeinschaft (DFG) "SPP1551—Ressourceneffiziente Konstruktionselemente" (2012–2017). Additional to the data from the literature different assumption for the LCA are made to simulate some effects and demonstrate the method. For the project roller bearings type NU206 are hard turned and deep rolled and compared to conventional ground and honed bearings regarding the lifetime.

The second part is a notched rotating bending part which is turned and deep rolled. The part is made of AISI 4140 (German grade 42CrMo4) and the dimensions are given in Fig. 3. Rotating bending tests are carried out on a 4-point-bending test rig type ROTABEND 400 by SincoTec. Three different specimen types are tested to identify the fatigue limit for each type. In a first step the fatigue limit for turned parts with no subsequent finishing operation is identified. In a second step the fatigue limit for turned and deep rolled specimens is identified. And in the final third step the specimen will be reduced in diameter and will be turned and deep rolled to achieve a similar fatigue limit as the just turned parts.

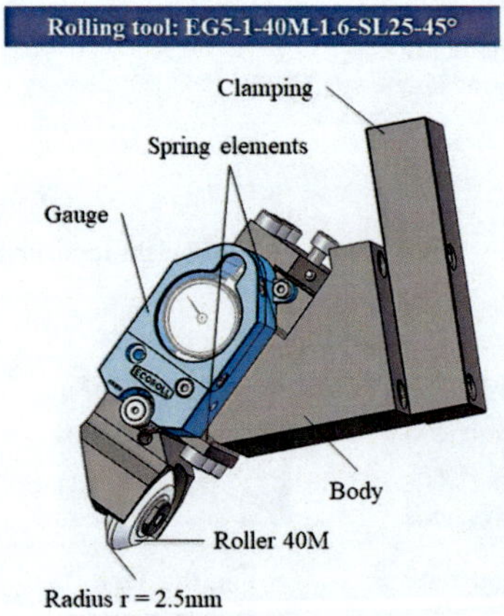

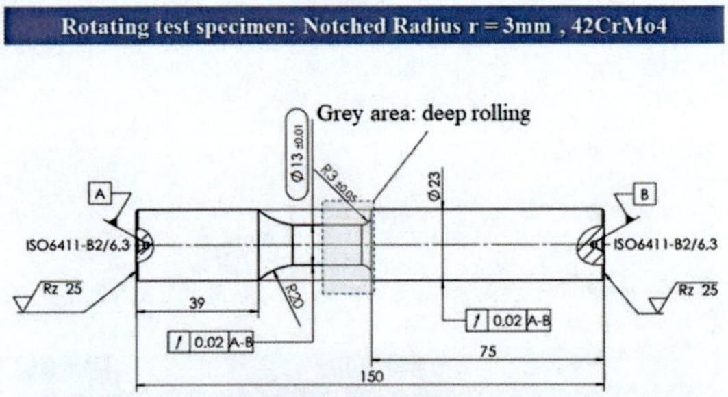

Type	Turning	Deep rolling	Diameter
1	X		13 mm
2	X	X	13 mm
3	x	X	< 13 mm

Fig. 3. Tool and part geometry for the deep rolling of the rotating bending specimen.

3 Rolling of Specimens

The highest loaded part in a roller bearing is the inner ring. Therefore, the inner rings of roller bearings type NU206 are hard turned and deep rolled on a Hembrug Slantbed Microturn 100, which is a high presicion lathe. For the deep rolling a hydrostatic rolling tool type HGx from ECOROLL AG Werkzeugtechnik is used, where x displays different tool diameter. To identify the optimum parameters for the process to improve the lifetime the pre-machining (hard turning), the rolling ball diameter, the rolling pressure and the overlap factor is varied. In summary the deep rolling process reduces the surface roughness from the turning operation by about 50%. The residual stresses and hardness are only defined by the deep rolling process.

Which means to generate a surface roughness of Rz < 1 µm the hard turning already needs to generate a roughness of Rz < 2 µm. With feed values of less than f = 0.1 mm this is achieved within the cited project. For the deep rolling process, Maiss identified the most important factor, regarding the surface quality, the overlap factor. In the experiments an overlap factor of more than 70% generates a good surface, as summarized in Fig. 4, left [16, 18].

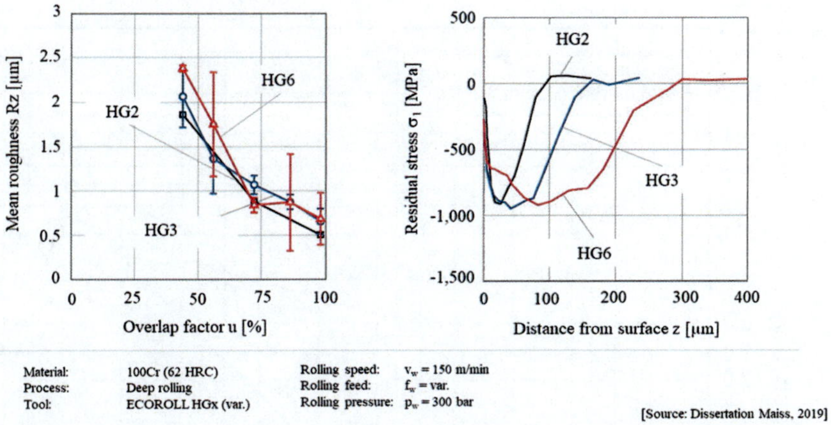

Fig. 4. Effects of deep rolling on surface roughness and residual stresses by Maiss [16].

To induce the necessary compressive residual stresses, it is found for the bearings to use a ball diameter of $d_k = 6$ mm, which is respectively a tool type HG6. As it is known from the hertzian contact, larger contact diameters are inducing stresses into greater depth. The second factor is the rolling pressure, which affects the value of the maximum compressive stress. The overlap factor also changes the residual stress state, but this effect is less than for rolling pressure and ball diameter.

For the rotating bending tests the specimens are all deep rolled with a mechanical deep rolling tool type EG5-40M from ECOROLL AG Werkzeugtechnik. In contrast to the machining of the bearings, this tool has no additional hydraulic unit which consumes additional energy. By using this kind of tool, both possible tool types, hydrostatic and mechanical, are considered within this study. This tool consists of a roller with a diameter of d = 40 mm and a radius of r = 2.5 mm. The roller is pressed to the surface in an angle of 45°. The force is applied by a spring which is applied by the machine tool. With this kind of tool, the notch radius of r = 3 mm (R3) can be machined with a constant feed value and indentation. The process parameters are chosen to be an indentation to apply a rolling force of $F_W = 2000$ N for the cylindrical area of the part. The feed value is $f_W = 0.1$ mm and the number of revolutions of $n = 600\ \text{min}-^1$, which means a rolling speed of $v_W = 24.5 - 43.4$ m/min, depending on the diameter within the radius.

4 Increasing the Lifetime of Mechanically Loaded Parts

The effect of deep rolling on the lifetime of mechanically loaded parts can be summarized as a training effect of the material. That means the induced residual stresses apply a counter load to the actual loads during the parts lifetime.

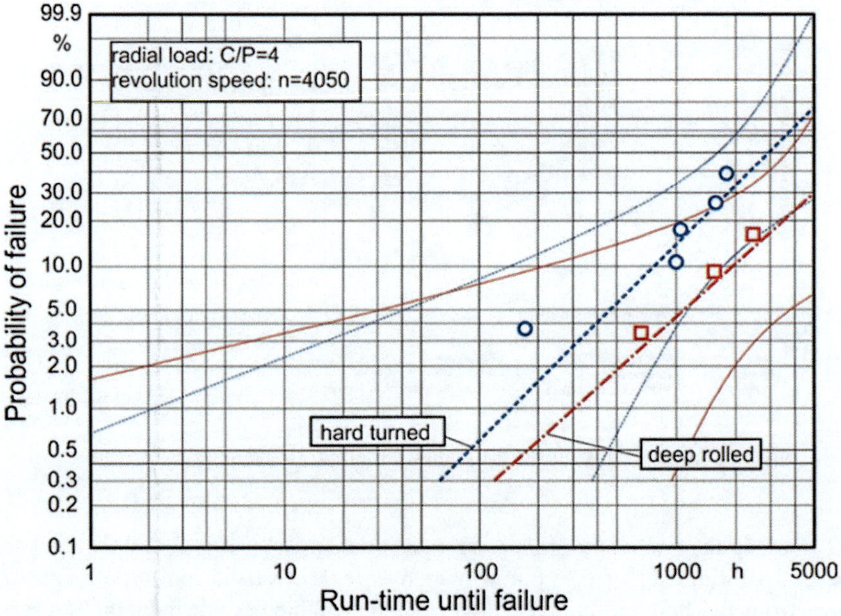

Fig. 5. Increased fatigue life of deep rolled roller bearings NU206 [18].

As described in the introduction, by inducing more compressive stresses as in the stable phase would be reached during the manufacturing, as it is done by deep rolling, the bearing lifetime can be expanded. The results of the cited research project indicate a lifetime increasement by the factor of 2.5, which means the deep rolled bearings reach 150% of the lifetime of conventional bearings. The Weibull diagram in Fig. 5 shows the results [16–18].

For the rotating bending tests the parts are tested to identify the load for the fatigue limit. At the time this paper is written, the experiments are still running. From previous experiments an increase of about 30–40% can be expected. The actual data will be given at the conference.

5 Concept for Resource Efficient Lightweight Design

To reduce the material weight and therefore the CO_2 emissions, the approach of Fig. 2 will be applied in the following of this paper. The idea is to use the deep rolling process to strengthen parts with less material to achieve the required fatigue limit. In the example of the roller bearings, this can be done by using the experimental data given in the previous sections and the well-known bearing lifetime calculations. The experiments show, that the fatigue increasement factor for this roller bearings is 2.5. To compare the lifetimes a lifetime of a large and a small bearing is used, respectively $L_{10,L}$ and $L_{10,S}$. Using the lifetime increasement factor.

$$L_{10,L} = 2.5 \times L_{10,S} \tag{1}$$

Using the given basic dynamic load rating values C_i from the bearing manufacturers, Eq. (1) can be transformed to.

$$(C_L/P_L)^p = 2.5 \times (C_S/P_S)^p \tag{2}$$

where for the NU206 from the experiments $p = 10/3$ and $P_L = P_S$, because the bearings is supposed to handle the same loads. Transforming Eq. (2) the new smaller bearing needs a bearing constant of C_S calculated by.

$$C_S = C_L/2.5^{1/p} \tag{3}$$

The bearing basic dynamic load rating C for the NU206 is given by $C_L = 46{,}000$ N. According to Eq. (3) the smaller bearing must have a $C_S = 34{,}944$ N, which is given for a bearing type NU205.

This concept demonstrates one possible solution to apply a lightweight design for roller bearings by using deep rolling. It is based on assumptions which have to be approved in the future. One of these assumptions is, that the NU205 will have the same lifetime increasement factor as the NU206. Also, it is known, that the applied calculation method simplifies the bearing calculation. However, at this point, this concept is applied to give an idea how the lightweight design concept can be applied within the design phase. A similar approach will be used for the rotating bending tests. In contrast to the bearing tests, these results will be proven by actual experiments in the future.

6 Evaluation of the Resource Efficiency

To evaluate the material reduction, the material used for the lighter NU205 bearings is compared to the material used for the heavier NU206 bearings. To evaluate the effect of deep rolling on CO_2 reduction, it would be necessary to detect the energy consumption in the manufacturing process and measure all input and output resources (e. g. coolant or tool wear particles). So far, literature data does not allow this detailed evaluation. Therefore, the reduced amount of material will be taken into account at this point. The process chains for the production of both bearings are depicted in Fig. 6. It is visible that both process chains are similar with hard turning and deep rolling processes substituting the finishing operations grinding and honing. Both production chains start with full AISI52100 bars which are machined to size using turning operations. The input weight is estimated by using full bars with the outer diameters of the inner and outer ring as well as the rolling elements.

Fig. 6. Process chains for the bearing production.

The NU206 has a total mass of $m_{NU206} = 200$ g, compared to a mass of $m_{NU205} = 132$ g for the bearing NU205. This means a reduced amount of bearing steel of about 33%. This reduced amount of material leads to an increased resource efficiency. To determine the CO_2-emissions connected to the production of the different bearing sizes, the best method would be to perform a lifecycle assessment (LCA) of both production chains. In order to perform such an analysis, process knowledge for every production step (i.e. energy consumption, tool wear, auxiliary supplies) would be necessary. Because this knowledge does not publicly exist yet, only a careful estimation can be made. Considering that only one step is substituted, only the difference between the combinations grinding-honing and hard turning-deep rolling have to be considered. For these processes, the energy consumption was measured in similar magnitudes [16]. It can therefore be concluded that the weight reduction will also lead to a reduction of CO_2 emissions.

7 Summary and Outlook

This paper introduces a method to use surface engineering, in specific the use of residual stresses due to the mechanical surface treatment deep rolling, to evaluate a part weight reduction. This weight reduction is performed with the goal to reduce production related CO_2 greenhouse gas emission and material resources. Deep rolling is a finishing processes that deforms the surface and subsurface area of metallic parts. It is used to reduce the surface roughness and induce compressive residual stresses. Especially residual stresses can increase the fatigue limit of dynamically loaded parts, as for example roller bearings, shafts, or axes.

By deep rolling the parts, dimensions can be reduced without decreasing the lifetime of the actual part. Within this paper, literature data is used demonstrate the method and to identify the lifetime increasement by deep rolling for roller bearings. The discussed additional tests with rotating-bending-specimens are described to identify the lifetime increasement for rotating-bending-loads. The next step of this project will be to finish the fatigue tests of the rotating bending tests. To evaluate the described method, a roller bearing NU206 is replaced by a smaller bearing type NU205. Using deep rolling, both bearings can achieve the same lifetime. At the end of this paper the resource efficiency is estimated by the evaluation of the amount of material used for the parts.

An evaluation of the CO_2 footprint will be conducted in the following steps of this project. To perform this analysis, the same procedure will be conducted with the described rotating-bending-tests. The lifetime test will be conducted and a specimen with a reduced weight will be designed and tested. With the whole resource data on the manufacturing process, as e. g. energy, material or auxiliary supplies, a Life-cycle-analysis can be conducted with reasonable assumptions.

In the future it will be necessary to take additional setups into account. It would also be interesting to include the effect on surrounding parts, which also can be decreased in size.

References

1. Statista: Größte Autokonzerne nach Umsatz bis (2019). https://de.statista.com/statistik/daten/studie/160831/umfrage/umsatzstaerkste-autokonzerne-weltweit/. Accessed 13 Feb 2022
2. BMWi: Innovation durch Forschung – Erneuerbare Energien und Energieeffizienz: Projekt und Ergebnisse der Forschungsförderung 2015 (2016)
3. Raedt, H.-W., Wilke, F., Ernst, C.-S.: Initiative Massiver Leichtbau – Leichtbaupotenziale durch Massivumformung. ATZ – Automobiltechnische Zeitschrift **116**(3), 58–64 (2014)
4. Field, M., Kahles, J.: Review of surface integrity of machined components. CIRP Ann. **20**(2), 150–163 (1972)
5. Breidenstein, B.: Oberflächen und Randzonen hoch belasteter Bauteile. Leibniz Universität Hannover, Habilitationsschrift (2011)
6. Wohlfahrt, H.: Mechanische Oberflächenbehandlungen: Grundlage-Bauteileigenschaften-Anwendungen. Wiley (2000)

7. Arola, D., Williams, C.: Estimating the fatigue stress concentration factor of machined surfaces. Int. J. Fatigue **24**(9), 923–930 (2002)
8. Meyer, K., Denkena, B., Breidenstein, B., Abrão, A.: Influence of residual stress depth distribution on lifecycle behaviour of AISI4140. Procedia CIRP **87**, 450–455 (2020)
9. Meyer, K.: Lastangepasste Randzonenfunktionalisierung, PhD dissertation, Leibniz Universität Hannover (2021)
10. Voskamp, A.P.: Microstructural changes during rolling contact fatigue. PhD Dissertation, TU Delft (1996)
11. Voskamp, A.P., Mittemeijer, E.J.: State of residual stress induced by cyclic rolling contact loading. Mater. Sci. Technol. **13**(5) (1997)
12. Hacke, B., Radnai, B., Hinkelmann, K.: Berücksichtigung von Betriebszuständen, Sonderereignissen und Überlasten bei der Berechnung der Wälzlagerlebensdauer von Windkraftanalagen und Großgetrieben. Abschlussbericht, FVA-Forschungsheft, Vol. AiF-Nr. 15227 N (2011)
13. Brinksmeier, E., Klocke, F., Lucca, D., Sölter, J., Meyer, D.: Process signatures – A new approach to solve the inverse surface integrity problem in machining processes. Procedia CIRP **13**, 429–434 (2014)
14. Jawahir, I., et al.: Surface integrity in material removal processes: Recent advances. CIRP Ann. **60**(2), 603–626 (2011)
15. Meyer, D., Kämmler, J.: Surface integrity of AISI 4140 after deep rolling with varied external and internal loads. Procedia CIRP **45**, 363–366 (2016)
16. Maiß, O.: Lebensdauererhöhung von Wälzlagern durch mechanische Bearbeitung. PhD Dissertation, Leibniz Universität Hannover (2019)
17. Neubauer, T.: Betriebs- und Lebensdauerverhalten hartgedrehter und festgewalzter Zylinderrollenlager. PhD Dissertation, Leibniz Universität Hannover (2016)
18. Denkena, B., Poll, G., Maiß, O., Pape, F., Neubauer, T.: Enhanced boundary zone rolling contact fatigue strength through hybrid machining by hard turn-rolling. Proceedings of the Bearing World Conference, pp. 87–102. FVA, Hannover (2016)

Bio-based Material

Approaching a Smart, and Sustainable Interior for Future Mobility Solutions

Torben Seemann[1,1(✉)], Claudia Burgold[2], Sergey Stepanov[3], Marvin Christopher Vincenzo Omelan[4], and Sebastian Stegmüller[5]

[1] Fraunhofer Institute for Surface Engineering and Thin Films IST, Bienroder Weg 54E, 38108, Braunschweig, Germany
torben.seemann@ist.fraunhofer.de

[2] Fraunhofer Institute for Wood Research, Wilhelm-Klauditz-Institut WKI, Bienroder Weg 54E, 38108, Braunschweig, Germany
claudia.burgold@wki.fraunhofer.de

[3] Fraunhofer Institute for Manufacturing Technology and Advanced Materials IFAM, Bienroder Weg 54E, 38108, Bremen, Germany
sergey.stepanov@ifam.fraunhofer.de

[4] Fraunhofer Institute for Surface Engineering and Thin Films IST, Bienroder Weg 54E, 38108, Braunschweig, Germany
marvin.omelan@ist.fraunhofer.de

[5] Fraunhofer Institute for Industrial Engineering IAO, Nobelstraße 12, 70569, Stuttgart, Germany
Sebastian.Stegmueller@iao.fraunhofer.de

Abstract. Electrification and automated driving functions will have an enormous influence on automobiles of future generations, resulting in particular in an increased focus on the interior. New operator concepts as well as free time gained through autonomous driving will require new solutions and offerings for vehicle interiors. In addition, increased requirements for sustainability and the reduction or neutrality of CO_2 emissions of materials, manufacturing processes and final products will have a massive impact on the design of future vehicles. A modular interior is thought off for different usage scenarios for one vehicle: the mobile office for the daily trip to work or to a meeting, a bulk purchase or group excursion or the transport of goods in between times. Competencies in renewable raw materials, their processing and component functionalization are bundled for the design of future passenger car interiors and light commercial vehicles. The aim is to develop quickly replaceable components which, on the one hand, allow the interior to be adapted to its respective purpose, having a long service life thanks to robust surfaces, and, on the other hand, guarantee repair and maintenance during the continued use of the vehicles in order to reduce vehicle downtimes. In this paper the development of an interior demonstrator for future mobility will be shown.

K. Dröder and T. Vietor (Eds.): *Future Automotive Production Conference 2022, Zukunftstechnologien für den multifunktionalen Leichtbau*, pp. 99–111, 2023.
https://doi.org/10.1007/978-3-658-39928-3_8

Keywords: Future mobility · Sustainable interior · Renewable materials · Functional components · Sustainable surface treatment · LCA

1 Overview

The results of the subproject "Future Interior" within the project "FutureFlexPro - Development of variant flexible and eco efficient system components for future vehicle generations in the sense of a holistic circular economy" will be shown. Within this project an interdisciplinary approach combining the know how of several Fraunhofer Institutes resulted in the concept and manufacturing of a sustainable and function integrated door panel for future cars and the valuation of the corresponding processes and technologies. Work was starting from the design of the panel, the selection and processing of bio-based materials, their functionalization such as self-cleaning and dirt-repellant surfaces, metallization, and sensor integration. Furthermore, a Life Cycle Assessment (LCA) of the surface pretreatment processes prior to the material joining was carried out, showing the atmospheric pressure plasma technique to have a much lower environmental impact than conventional methods. The panel itself and its features for the interaction between occupants and vehicle according smart surfaces will be shown.

Within the project a survey has been conducted about the acceptance of sustainable and recycled materials in an automotive interior in the three automotive lead markets USA, China and Germany.

2 Objectives

The overriding goal of our joint work is the conception and exemplary implementation of sustainable, modular components for future mobility vehicles, taking into account complete life cycles from the manufacturing, use and disposal phases as well as the evaluation of ecological aspects and business model considerations. Current developments can be seen in [1], Appropriate modular concepts must be examined with regard to their economic competitiveness with classic solutions for highly efficient, unit-oriented production in order to ultimately be able to obtain a market and application opportunity. In particular, high potentials for the effective use of components are to be expected through the decoupling of stress and lifetime of these components. Derived from this, the question of the optimal design arises in the area of tension between economies of scale and modular design. This must be carried out taking into account various utilization and number-of-units scenarios.

Fig. 1. Design of a door panel used for a sustainable, modular interior demonstrator.

3 Results

3.1 Demonstrator manufacturing

In the interior sub-project, two different sets of demonstrators were built, one veneer-based, one nature fiber reinforced plastic based. The general design can be seen in Fig. 1.

The veneer-based, functionalized demonstrator components were successfully produced in a sandwich design with targeted surface functionality and sensor integration using a vacuum process. The demonstrators are made of lightweight particle foam and birch veneer plywood, one example can be seen in Fig. 2. According to the current state of the art, veneer in interiors is coated with several layers of varnish to meet the requirements of the mobility sector. However, the optic and above all the haptic properties of the natural material wood are lost in the process. One aim of these researches was therefore to use a functional coating that preserves the naturalness of wood and at the same time has dirt-repellent properties. Plasma-enhanced chemical vapor deposition (PECVD) processes were used to achieve functional surfaces to adjust the hydrophobicity and dirt repellency of the top layers made of walnut and maple veneers, as well as vegan leather in various combinations. Here, a set-up at atmospheric pressure was used, where nitrogen was enriched with suitable film forming precursors optimizing the conditions of the plasma process [2–7]. The lamination of the temperature sensor initially serves as a first feasibility demonstration for the integration of a sensor into a veneer-based material, whereby neither the sensor nor the material should be damaged. For demonstration purposes it is conceivable to connect the sensor to temperature indicating LEDs or to a surface heating system.

In further investigations, other sensors are of course possible, as well as a wide variety of user interfaces for the passenger. For the integration of the sensor inserts, bonding tests were carried out beforehand to determine both the material of the substrate films and the adhesive. Both polyethylene terephthalate (PET) and polyimide (PI) have proven to be suitable substrate films. Coating the sensors with PI film also protects the sensors in the plywood. A 1-component polyurethane has proved suitable as an adhesive between the veneer and the substrate films. Pre-activation of

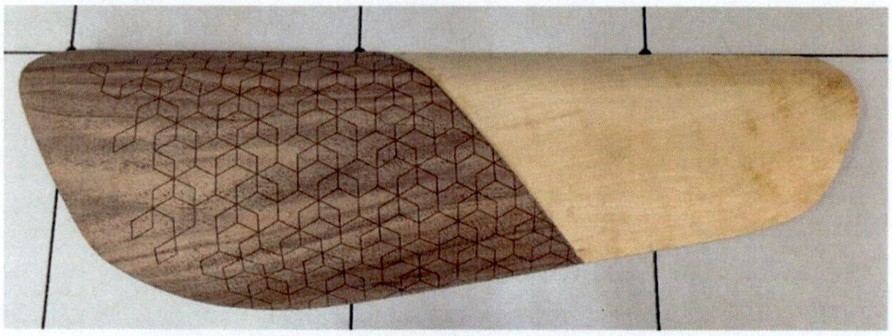

Fig. 2. A veneer based door panel with laser structured walnut top layer before applying the vegan leather.

the substrate foils by an atmospheric pressure plasma using pure nitrogen led to an improvement in the adhesion properties of the foil-veneer composite.

Finally, a thin-film-based temperature sensor was successfully developed as part of the project. Suitable polymer-based substrate materials were investigated and evaluated, and adapted cleaning, coating and microstructuring processes were developed for manufacturing the sensor insert. By means of several design variants, the electrical resistance range of the sensors could be adapted to the required range. The final electrical contacting and characterization of the thermoresistive sensor behavior showed

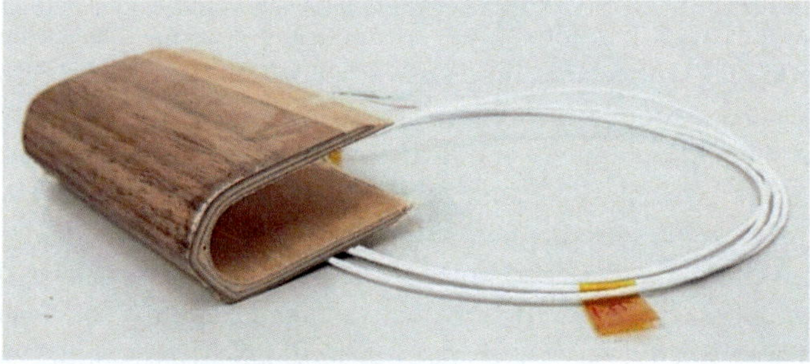

Fig. 3. Veneer sample with an integrated sensor.

the expected linear T-R dependence. Also, the forming process had no negative influence on the functionality of the sensors, see Fig. 3.

One challenge, however, was the thickness of the contact points and cables of the sensor integrated in the veneer layer structure, which led to clearly visible unevenness in the component, see Fig. 3. This can only be minimized by flatter contact points. In order to reduce the influence of the cables, a solution was devised in which the cables are installed on the front side of the component and are routed away hidden in a clasp, see Fig. 1.

As second demonstrator, a textile-based, function-integrated interior component was developed. The component consists of a flax fiber fabric, conductive silver-coated polyamide yarns, embedded in a plastic matrix of partially bio-based epoxy resin, designed and manufactured at HOFZET of Fraunhofer WKI, see Fig. 4. The conductive yarns can be embroidered on before the fabric is processed into a component or directly woven in during fabric manufacture. In the second case, conductive yarns were inserted into the natural fiber fabric at regular, defined intervals of 5 or 10 mm

Fig. 4. A door panel based made from bio-based materials at Fraunhofer WKI, HOFZET enabling integrated touch functions.

respectively. The fibers are the basis of the integrated sensor systems. A touch function with simple gestures could be build up without any additional sensoring system, see Sect. 3.2.

The challenge in the project was to contact the embedded conductive fibers without impairing the function or allowing the contact to be perceived on the visible side. The chosen solution also had to ensure that the fine fibers were contacted in a reproducible manner. Access to the fibers can be guaranteed even after embedding in the resin matrix via material accumulation on the underside of the component.

3.2 Functionalization of the Demonstrators

With regard to the functional coating of the veneer surfaces, it was possible to evaluate coatings that are stable over the long term while retaining the haptic and optical properties in order to reduce soiling. Particularly noteworthy are subsequent veneer reshaping processes, which have no negative influence on the surface coating.

Analytical verification was carried out by determining the surface energies determined from the contact angels of water and diiodomethane, see Fig. 6.

Targeted bio-based coatings using atmospheric pressure plasma and time-dependent measurements of coating stability by determining the contact angle as well as the

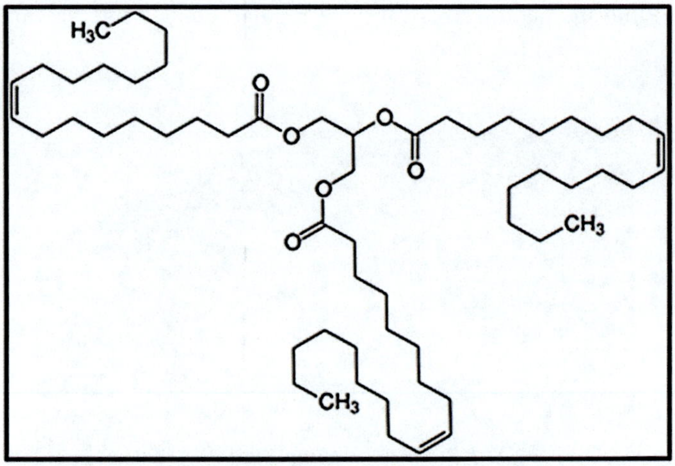

Fig. 5. Exemplified unsaturated fatty acid structure (glycerine trioleate).

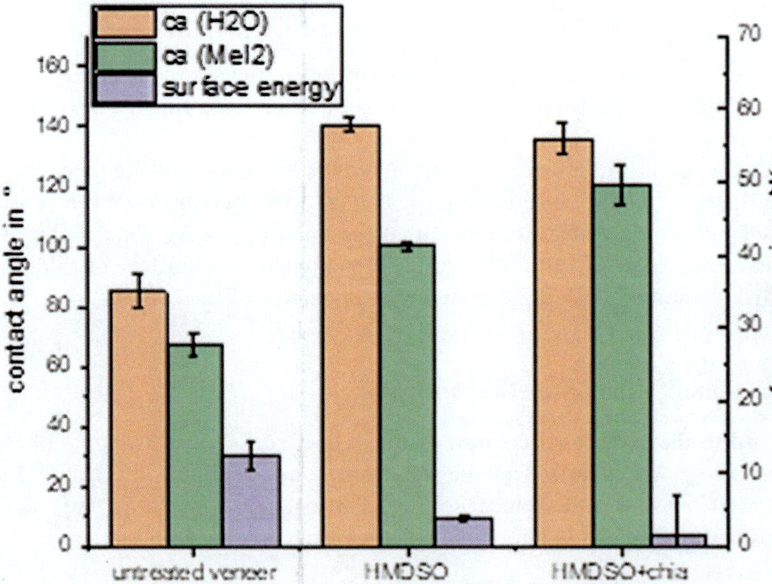

Fig. 6. Contact angles of water (H_2O) and diiodomethane (MeI_2) and surface energy of hexamethyldisiloxane (HMDSO)/chiaoil coated veneer in comparison to an untreated and a solely HMDSO coated wood substrate.

surface energy show a clear trend towards superhydrophobic surfaces. Especially, the use of unsaturated systems are of a great interest, due to the molecular double bond characteristics that offer a significantly higher polymerization rate. Here a careful control of energy input is required and thus also the radical formation (structural damage) of the sustainable coatings to the substrate is minimized. The structural building blocks of especially unsaturated bio-based oils are known and can be used for sustainable plasma polymerization [2–5]. An essential basic feature is the long and partly

Fig. 7. Repellency-tests on HMDSO/chiaoil coated walnut veneers by the use of fluorescent marker, coffee and pure iodine.

unsaturated hydrocarbon backbone, see Fig. 5. The higher the degree of unsaturation (e.g. amount of triple-unsaturated fatty acid), the more crosslinked network can be

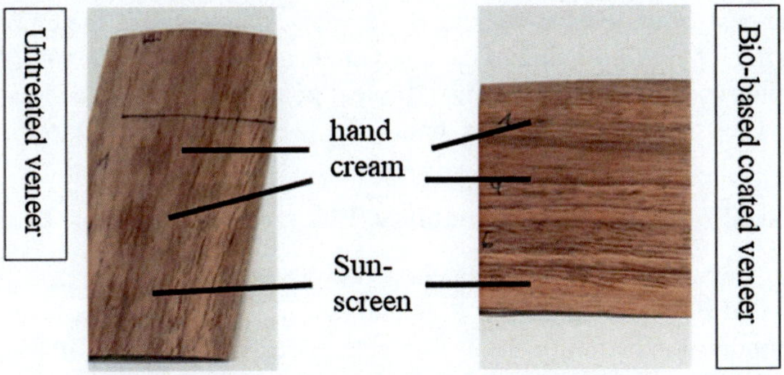

Fig. 8. Dirt-repellence tests of plasma-treated veneers based on hand creams and sunscreen.

formed by plasma-polymerization to achieve properties, such as hydrophobicity and dirt-repellency on the surface.

The hydrophobicity of the wood veneer can be significantly increased. Exemplary, a hydrophobicity test on a chia oil film on a walnut veneer was carried out, as shown in Fig. 7. All three solutions (fluorescence marker, coffee solution and iodine) draw instantly into the untreated wood material and shows a widened pollution. After plasma-enhanced deposition of chia oil the water-based marker and coffee solution repels from the surface easily. Also, the iodine repels from surface while maintaining clean veneers.

Cream resistance tests proved to be challenging, as very different results were generated, since differently tested hand creams and sunscreens have very different compositions (including perfumes), see Fig. 8.

In the development of a textile-based, functionalized demonstrator, proof of function was provided. With this demonstrator adjustment of ambient light as well as a heating for the passengers can be shown with turning the system on by swiping to the front, locking it against unwilling changes by swiping to the back and adjusting light or heat by holding the hand on the surface when turned on. The embroidered button can be used for switching back and forth between functions. One use case we see with this rather simple and intuitive gestures is the interaction with a level 5 autonomous vehicle giving passengers anywhere in the interior of the vehicle the chance to interact with it, e. g. opening the window, stopping the car or adjust ambient light and heating as we show in the demonstrator. These functions might be transferred to any component based on shown materials.

With this demonstrator the 5 mm intervals did not give any improvement regarding the resolution of the touch sensor as the signals could not clearly be separated. With 10 mm intervals the signals from the single fibers could be distinguished and described functions could be integrated well.

In the course of demonstrator production, a number of manufacturing challenges were identified, in particular the integration of contact points. In order to provide functional proof, the conductive yarns were first routed out of the infusion area. However, subsequent trimming of the component is only partially possible as a result. One goal for follow-up projects must be to introduce contacting possibilities in the weaving process or to develop a contacting process that accesses the embedded yarns directly through the reverse side of the components.

In addition to the previously defined project goal, the woven-in conductive threads can also be used as sensory elements that can be used to design a 3D touch surface.

3.3 Surface Pretreatment Sustainability (Life Cycle Assessment, LCA)

In order to investigate the environmental performance of the surface pretreatment techniques that might be employed in manufacturing interior components (e. g. prior adhesive bonding or paining), the Life Cycle Assessment (LCA) of the wet-chemical cleaning with isopropanol and the dry atmospheric pressure plasma treatment of polypropylene substrates was carried out using Umberto® software (ifu Hamburg) [8]. Thereby, according to the LCA methodology in line with ISO 14044 [9], the system boundaries included an adequate surface pretreatment of a 1 m^2 of polypropylene substrates (functional unit), whereas the Global Warming Potential (GWP in Kg CO_2-eq.) was considered as the midpoint impact category for the Life Cycle Impact Assessment (LCIA) of the pretreatment processes.

For the Life Cycle Inventory (LCI), the consumption of isopropanol and the paper wiping cloths amount needed for the desired treatment result were identified as decisive impact factors (input flows) for the wet-chemical cleaning and determined experimentally, with regard to the functional unit, see Table 1. For the atmospheric

Table 1. Functional unit and key impact factors (input flows) used in LCIA of surface pretreatment techniques.

Functional unit	Surface pretreatment of 1 m^2 of polypropylene substrates			
Pretreatment technique	**Input flow**	**Consumption**	**GWP [kg CO_2-Eq.]**	
Wet-chemical cleaning	Isopropanol	380 g	0.323	0.624
	Paper wiping cloths	280 g	0.301	
Atmospheric pressure plasma treatment	Compressed air	0.041 m^3	0.004	0.005
	Electricity	0.02 kWh	0.001	

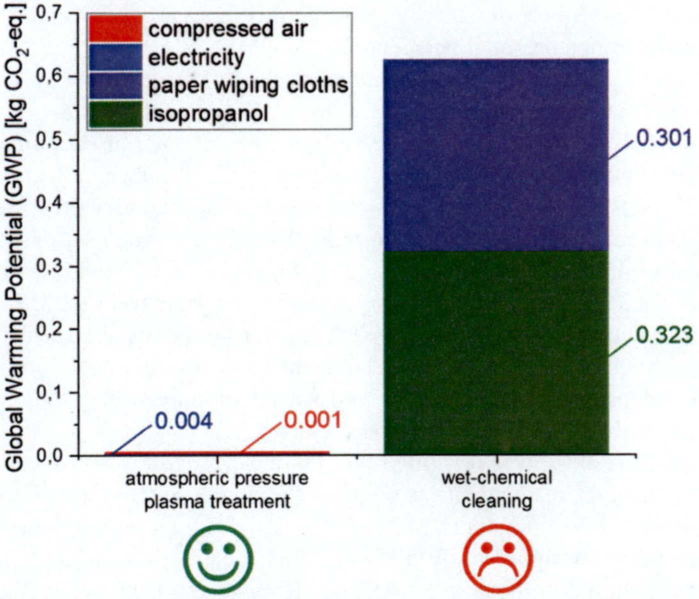

Fig. 9. Contribution of key impact factors to GWP of: (i) dry atmospheric pressure plasma treatment and (ii) wet-chemical cleaning.

pressure plasma treatment, the actual consumption of the process gas (compressed air) and the electricity consumption were found to be significant and established experimentally, also with regard to the functional unit and the plasma process parameters, see Table 1.

Basing on the ecoinvent database [10], both the total GWP and the GWP of each single input flow identified above were derived to characterize the environmental impact of the considered pretreatment techniques. The results of the performed LCIA are shown in the bar diagram in Fig. 9 and listed in Table 1.

Comparing the investigated surface treatment methods, the GWP of the dry atmospheric pressure plasma processing is only about 1% of the GWP value of the wet-chemical cleaning. This indicates a particular environmental friendliness and sustainability of the plasma processing in comparison to the conventional techniques.

3.4 Acceptance Studies

Beneath the assessment of the technical realization and the life cycle impact of new interior material it is crucial for evaluating a possible market success to understand if those materials will be accepted by users and if they are willing to pay for those innovations. For gathering this information and to build up a holistic picture of the introduced innovations, an international user survey was conducted to reflect the technical research from the project from a users' perspective. The questionnaire was divided into different parts: A first part contained a few questions that allow building different target groups, such as socio-demographic or character-related segments. The second part related to the comparison of different material groups and the last part was designed to ask questions that help to identify how users perceive the different material groups from a personal perspective.

The acceptance study is dedicated to the analysis of target group-specific requirements for the material world inside a vehicle. The transformation of production processes toward the use of sustainable and recyclable materials requires a comprehensive understanding of how different materials and substances affect and are perceived by the consumers. Their attitudes and expectations can vary greatly and depend on personality, nationality or generation type. To ensure a broad set of different user types, a total of 4544 subjects participated in the study, of which 1512 subjects were from China, 1503 subjects from Germany and 1529 subjects from the USA. The analysis provides valuable and target-group-specific starting points as to which properties need to be addressed when used in vehicle interiors. Furthermore, the results give insights about those aspects that have to be focused on more strongly in order to communicate the attributes of the material more clearly.

As a general result it was identified that the material groups wood and leather are perceived by the most respondents as being of the highest quality. Plastics and textiles are perceived by the respondents to be of the lowest quality. In Germany, wood is perceived as being of significantly higher quality. Leather, on the other hand, is the most favored material within the USA; this effect is also statistically significant. In general, it can be seen that at the Chinese market, the younger generations, as well as users with dominant or stimulated characters are more polarized regarding to the evaluation of the presented materials. Within the Chinese sample, the material group metal is significantly more strongly associated with high quality. The traditionalists perceive leather as being of significantly higher quality, while the younger generations in particular also rate alternative materials such as stone or composite material better.

The measurement of the semantic differential that was calculated based on the results, shows material-dependent differences:

- Wood is perceived similarly by all subgroups and does not show major valuation differences. The material is perceived as attractive, natural, warm and of high quality.
- Textile is rated more in the direction of cheap and artificial, but convinces with comfort and warmth. The German respondents perceive the material as less attractive, safe and stable. There are no significant differences in the assessment of the generations and the personal character of the participants.
- Leather-like materials are perceived across the board as being expensive and of high quality. Regarding the generations a very homogeneous evaluation shows up. If one compares the personality types, the dominance types in particular favor this material group and evaluate their characteristics much more positively.
- Composite material receives a rather moderate rating, which can only be assigned to one direction to a limited extent. It is striking that the material is not perceived as dominant natural. For this group of materials, knowledge penetration within society is rather low. As a result, subjects may have had problems evaluating the material group according to its properties, since no (conscious) empirical values could be used. This effect is particularly noticeable in the German market and among traditionalists.
- Stone is rated somewhat more moderately. Here, the characteristics high value, stable, expensive and cold dominate. The country comparison shows that the material is perceived as significantly more attractive for the American market. Germans, on the other hand, tend to perceive the material as uncomfortable. Younger generations, as well as the dominant and stimulant types, tend to give this material positive attributes.
- Metal is generally perceived as cool, stable and uncomfortable. The Chinese market in particular is convinced of its value and perceives it as significantly more attractive. The German respondents also see this material as more harmful to the environment compared with the respondents of the other two countries.
- If we look at the generations, we see a rather heterogeneous picture that varies only in the strength of the tendency. The stimulus and dominance types rate this group of materials much more positively.
- Plastic is perceived by the interviewees as rather moderate across the board and shows hardly any clear characteristics in the direction of an attribute. The material is evaluated as artificial and rather favorable, but at the same time also in the direction of stable and safe. While there are clear differences between the countries, the material is rated rather similarly across the generations and by the types and again varies only in the strength of the tendency, while the direction of the rating remains identical.

Some of the results were displayed on a material wall at the IAA Mobility 2021 in Munich. On the wall three different generations-related material and color concepts showing a mixture of sustainable and high-quality materials were exhibited. Real materials as developed and used at the FutureFlexPro-project were shown at the wall what allowed a better valuation of the materials by the visitors as during the picture-based online survey. The result is relatively clear: each generation perceives the materials differently - and thus poses special challenges for the designers and developers of the vehicle cabins of the future. Therefore, not only the material itself and its sustainable footprint must be considered in applied development processes, but these aspects must be evaluated in terms of influence and acceptance of specific target groups to ensure innovation success.

4 Conclusion and Outlook

Within the presented project the manufacturing and functionalization of a bio based door panel as example for a future interior component could be shown successfully. The production of wood-based demonstrators could be realized well in the vacuum process. However, until now several steps have been necessary to glue the individual components together. Further work is therefore needed to investigate how the manufacturing process can be optimized.

The effect of the coating of a vegetable oil on wood veneers by a Dielectric Barrier Discharge (DBD) based plasma process and the corresponding surface characteristics were presented. Based on the deposition of bio-based chia oil coatings a process was developed to improve the hydrophobicity on the sustainable surfaces. With optimized parameters, a delayed penetration time of water into the veneer surface was reached. Thus, regarding the cream stained surfaces, dirt-reducing coatings on the basis of a HMDSO/chia oil system can be achieved.

Furthermore, a Life Cycle Assessment (LCA) of the surface pretreatment processes prior to the material joining was carried out, showing the atmospheric pressure plasma technique to have a much lower environmental impact than conventional methods. In particular, this indicates a high potential of the plasma processing for the environmental friendliness and sustainability of the target future mobility solutions.

Related to the consumer survey, clear differences between generations and countries about preferred materials could be identified. This shows that also the use cases and target groups of vehicle projects should have been to be considered during the selection of sustainable materials, as it will influence the acceptance and willingness to pay of possible solutions. Based on the IAA experiment it can be suggested that further consumer research should also be done on the base of showing and valuating real materials. Those results could be compared to the ones of the presented international survey to extract strong recommendations.

Based on the results further work will be carried out in order to show a quick replacement of interior panels. These solutions might support quick repair for shared mobility which will support a high utilization time of vehicles. On the other hand, corresponding solutions will enable a cost-effective modernization of the interior. Main focus of the work will be mounts and integrated electronics.

Acknowledgements. The research with regard to variant flexible and eco efficient system components for future vehicle generations based on renewable materials (FutureFlexPro) was funded by the Federal Ministry of Education and Research (BMBF) (ref. L1FHG42421).

References

1. Möller, T. Schneiderbauer, T., Garms, F., Gläfke, A., Köster, N., Stegmüller, S., Kern, M. Werner, M., Bobka, K.: The future of interior in automotive, study, Studie. McKinsey & Company, Fraunhofer IAO. September (2021)
2. Rus, A.Z.M.: Polymers from renewable materials. Sci. Prog. **93**, 285–300 (2010)
3. Adekunle, K.: A review of vegetable oil-based polymers: Synthesis and applications. Open J. Polym. Chem. **5**, 34–40 (2015)
4. Mertens, J., et al.: Atmospheric pressure plasma polymerization of organics: effect of the presence and position of double bonds on polymerization mechanisms, plasma stability and coating chemistry. Thin Solid Films **671**, 64–76 (2019)
5. Gerchman, D., Bones, B., Pereira, M.B., Takimi, A.S.: Thin film deposition by plasma polymerization using d-limonene as a renewable precursor. Prog. Org. Coat. **129**, 133–139 (2019)
6. Herrmann, A., Lachmann, K., Fischer, K., Kovac, J., Thomas, M.: Area-selective epoxy coatings by DBD-PECVD in 3D cavities for protein coupling. Surface Innovations **3**(4), 206–214 (2015)
7. Vohrer, U., Bosse, R., Hoder, T., Černák, M., Thomas, M.: Reinigen, Funktionalisieren und Beschichten. Vak. Forsch. Prax. **28**, 26–32 (2016)
8. https://www.ifu.com/umberto/
9. ISO 14040:2006, Environmental management - Life cycle assessment - Principles and framework. Technical report, ISO- International Organization for Standardization, (2006)
10. https://ecoinvent.org/the-ecoinvent-database/

Fast Curing Biobased Epoxy Hardener for RTM Applications

Stefan Friebel[1(✉)], Ole Hansen[1], and Jens Lüttke[2]

[1] Fraunhofer-Institut für Holzforschung, Bienroder Weg, 54 E, 38108 Braunschweig, Germany
{stefan.friebel,ole.hansen}@wki.fraunhofer.de

[2] HOBUM Oleochemicals GmbH, Konsul-Ritter-Straße, 10, 21079 Hamburg, Germany
info@hobum.de

Abstract. Efficient lightweight solutions becoming increasingly important in the automotive industry. Since the trend of using sustainable electric engine drives is ongoing, the CO_2 impact of car components is becoming more important for the overall aim to produce a CO_2 neutral mobility until 2050.

So far, for automotive components almost exclusively carbon fibre reinforced plastics have been used in lightweight construction. Natural fibres offer an ecological alternative for non- or semi-structural car body parts. They exhibit, however, lower stiffness and strength than carbon fibres, but mechanical properties are sufficient for many applications for car body parts. Due to their naturally grown structure, natural fibres dampen sound and vibrations better. Their lower tendency to splinter can help reduce the risk of injury in the event of an accident. In addition, they do not cause skin irritation during processing.

The overall aim of the project is the development of a sustainable biosourced natural fibre reinforced epoxide for a car door. The research approach is a fast curing bio-sourced epoxy system for RTM (Resin Transfer Moulding) applications with a glass transition temperature>100 °C of the natural fibre reinforced composite. Due to its chemical structure, the bio-based epoxy resin has a tough elastic behaviour, which could offer advantages in the crash test. Among the used chemistry for the bio-sourced material, the kinetics of the bio-sourced resin will be considered and brought into relation of the whole production process. Additionally, data of dynamic thermomechanical analysis of the material will be presented.

Keywords: Lightweight design · Bio-sourced materials · Exterior · Bio-epoxy · Natural fibre

1 Introduction

Renewable raw materials, i.e. plants that are industrially cultivated for use as raw materials outside the food industry, have a potentially advantageous carbon footprint due to CO_2 fixation in their growth phase. For example, on average, only about 45 MJ

K. Dröder and T. Vietor (Eds.): *Future Automotive Production Conference 2022*, Zukunftstechnologien für den multifunktionalen Leichtbau, pp. 112–126, 2023.
https://doi.org/10.1007/978-3-658-39928-3_9

is required for one kg of natural fibre yarn [1]. In omparison carbon fibres require between 200 MJ/kg [2] up to 340 MJ/kg [3], i.e. 5–7 time more energy. Natural fibres, such as flax or hemp, have some advantageous properties, such as a lower tendency to splinter or a low density. Their mechanical properties are significantly lower in terms of elastic modulus and tear strength than those of carbon fibres, but are sufficient high for many applications in combination with epoxy resin [4–8]. To consistently continue the use of renewable raw materials, the use of a bio-based epoxy resin or hardener is therefore indispensable. Resins of epoxy systems comprise of molecules having an oxiran ring in their structure. These oxiran rings are thermodynamically instable and can be ring opened by acidic and basic reactants. The position of the oxiran ring in the molecule determines the reactivity of the ring opening reaction. Common epoxy resins which can be cured at room temperature are mainly glycidyl ethers of alcohols and the oxiran ring is at the edges of the molecule and thus easily accessible for reactants [9]. Oxiran rings in epoxidized vegetable oils are far less reactive since the rings are in the middle of a hydrocarbon chain of the triglyceride molecule and thus sterically hindered. Epoxidized vegetable oils therefore are typically cross-linked at high temperatures (120–180 °C) and referred to as thermosets [10] and [11]. Bio-epoxy resins are scarce products on the market. Commercially available are bisphenol-A-diglycidylether resins (BADGE) with a BIO-C (carbon origin from renewable resources) content of~28% originating from the reaction of mineral based bisphenol A with Bio-epichlorhydrin derived from Bio-glycerol originating from Bio-diesel production. Commercially available is also the glycidyl ether of Bio-glycerol. At present the BIO-C content can be measured by radiocarbon measurement (C14 Method ASTM D6866). The BIO-C content varies from 28 to 53%. Suppliers of bio-based epoxy resins are for example: Sicomin Epoxy Systems, Sika Advanced Resins. Resoltech, Entropy Resins, Inc, Spolchemie and AEP Polymers SRL.

Oxiran rings can be opened by various functional groups. Commercially available amines with a measurable BIO-C content are practically not available. Thus, polyamino amides, the reactants of amines with fatty acids are the typical hardeners of first choice, when looking for hardeners with a BIO-C content. Polyamino amides comprise of pre-polymers made of polyamines and polyvalent fatty acids. In their work, Ramon compare a large number of bio based systems with regard to their applicability as the matrix of a composite material in aviation and show the suitability of some solutions. In principle, however, the systems considered show a high scattering of the investigated characteristic values and no or only very limited commercial availability [11]. Mustapha's work gives an overview of vegetable oil-based epoxy resins combination with bio-based hardeners. Besides commercial availability, higher mechanical properties are desirable. Furthermore, he shows a high variability of the glass transition temperature [12].

Various studies show that glass transition temperatures well above 100 °C are possible, but this requires a curing time of 1 h or more [13–15]. This is in discrepancy when fast curing cycles (<20 min) are necessary in the course of series production using resin transfer moulding. In addition to the cycle times, energy-efficient process control is also a decisive factor here, especially nowadays. The impregnation of the semi-finished products in the cavity is a critical point, because long impregnation and curing times are associated with high costs. Heating or using fast curing resin systems offer the possibility to reduce the infusion time [16]. In this context, the viscosity of the resin also plays an important role for sufficient impregnation within given production times. In order to ensure optimal resin distribution, it should have a low viscosity and the inflow and outflow of the resin should be well distributed [17]. The pressure acting on the resin and the permeability of the fibre semi-finished products also influence the thickness of the finished workpiece [18]. The aim of the work in the project "BioResinProcess" is therefore the development of a fast-curing, bio-based resin-hardener system with a glass transition temperature in the natural fibre composite>100 °C. The feasibility is to be demonstrated on an exterior part of a vehicle component produced by resin transfer moulding.

2 Materials and Methods

2.1 Materials

2.1.1 Natural Fibre Fabrics

The natural fibre fabric used is AmpliTex™ 5040 with a basis weight of 300 g/m^2 from Bcomp.

2.1.2 Bio-Epoxy Resins

Th1e conventionally available partially bio-based epoxy resin Greenpoxy 56 (GP56) from the company Sicomin is used. The mixing ratio to the developed hardener is adjusted stoichiometrically via the epoxy or amine number. For the investigation of the energy consumption, the hardener SZ 8525 from Sicomin was used and cured for 25 min at 80 °C according to the technical data sheet.

2.1.3 Bio-Hardeners

In this study, amine-, amide-, carboxylate- and anhydride groups were utilized. By careful selection of the reaction partners, polyamino amid hardeners for resin transfer moulding with an initial BIO-C content of 81% have been synthesized. These hardeners are not able to generate a T_G of 100 °C when reacted with the available BIO-C epoxy resins. Modification of the polyamino amide pre-polymer with adducts of mineral based epoxy resins of high functionality allow to overcome the intrinsic flexibility of the polyamino amides and the cured. This modification decreases the BIO-C content from initially 81 to 64%.

2.2 Methods

2.2.1 RTM Process

A pressure pot system is used for the production of the natural fibre reinforced plastics. This is shown schematically in Fig. 1. A plate cavity with the dimensions 400 × 400 mm is used as a mould. The plate thickness is variably adjustable, for these experiments it is 2 mm. The tool is shown in Fig. 2. The mould is heated via water-fed temperature control units.

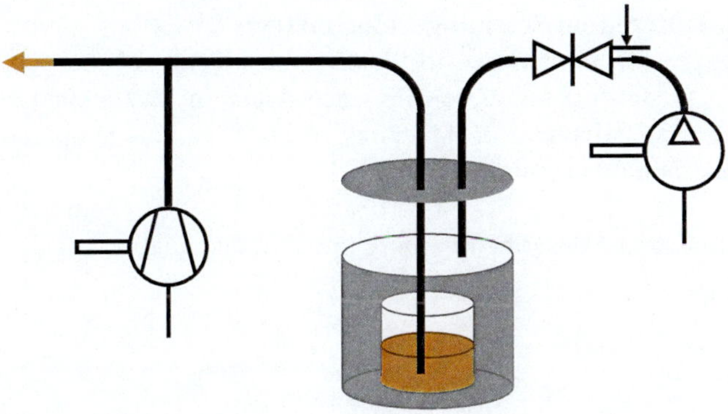

Fig. 1. Schematic representation of the pressure pot process

Fig. 2. Mould cavity with inlaid natural fabrics

After placing the fabrics in the preheated mould it's closed with a force of 120 kN. The cavity and pressure pot are then evacuated to a pressure of 2 mbar. This process is used for additional drying of the fibre so that the fibre moisture is below 2%. By applying pressure to the system, the resin is pressed into the cavity and cured. For the tests, the pressure was set between 3 and 9 bar. The mould temperature was varied between 105 and 115 °C. In the course of the energy recording, the entire process time was considered, including evacuation, drying and injection. This results in a total process time of 40 min for this sample.

2.2.2 Viscosity

As already stated the viscosity is important to control the RTM-process. The viscosity was measured by Kinexus Lab+ by Malvern Panalytical. After mixing the resin and the hardener a sample was placed between a preheated plate-plate Peltier-heated device. The measurement was carried out in oscillation mode with a 1% displacement and a gap of 1 mm.

2.2.3 DSC (Differential Scanning Calorimetry)

A Mettler DSC 3+/700 was used to identify the kinetics of the epoxy resins under investigation. Model-free-kinetic (MFK) according to Vyzovkin [19] was used to simulate the curing behaviour of the epoxy resins at various temperature. Model-free analysis is based on two assumptions:

a) The reaction can be described by only one kinetic equation for the degree of reaction α:

$$\frac{d\alpha}{dt} = A(\alpha)f(\alpha)e^{\left(\frac{-E(\alpha)}{RT}\right)}$$

where **E(α)** is the activation energy depending on the conversion **α**, and **A(α)** is the pre-exponential factor. R is the universal gas constant and T the temperature.

Activation energy E of chemical reaction is calculated without any assumption about reaction type, but pre-exponential factor A can be found only under assumption about reaction type.

b) The reaction rate at a constant value of conversion is only a function of temperature.

This method is especially common describing the kinetics of reactive resins, were due to viscosity change a simple n^{th}-order kinetic changes during the chemical curing into a more complex diffusion driven kinetic.

Additionally, an isothermal conversion of the epoxy system was measured by determining the integral of the enthalpy curve and calculation the time-dependent conversion by

$$\alpha = \frac{\Delta H_t}{\Delta H_0}$$

where ΔH_t is the enthalpy at time t and ΔH_0 is the total enthalpy of the reaction.

2.2.4 DMA (Dynamic Mechanical Analysis)

All measurements were carried out in a Tritec 2000 DMA by Mettler. The specimen geometry of the RTM-processed samples was $9 \times 5 \times 2\ mm^3$. All samples were measured with single-cantilever-modus. The displacement was 20 µm at a frequency of 1 Hz. A temperature range between 20–180 °C with a heating rate of 2 K / min. Was investigated. Evaluating the influence of the fibre on the DMA, the plates were prepared with windows in which only resin was cross-linked with the same thermal history compared to the NFC (Natural Fibre Composites) as seen in Fig. 3.

Fig. 3. Typical samples measured by DMA. Top: NFC; Bottom; Area of only resin in the NFC

2.2.5 Tensile Properties

The tensile properties were determined in accordance with DIN EN ISO 527-4. The Youngs modulus is calculated via the quotient of the difference in tensile stress and the difference in longitudinal by

$$E_t = \frac{\sigma_2 - \sigma_1}{\varepsilon_2 - \varepsilon_1}[MPa]$$

where: σ_2 is the stress measured at a strain value ε_2=0.0025 (0.25%); σ_1 is the stress measured at a strain value ε_1=0.0005 (0.05%).

The tensile strength σ_t as resistance to tensile load is calculated by

$$\sigma_t = \frac{F_{max}}{A_0}[MPa]$$

where: F_{max} is the maximum tensile force [N]; A_0 is the initial cross-sectional area of the specimen [mm^2].

2.2.6 Flexural Properties

The flexural properties were determined in accordance with DIN EN ISO 14125. The bending modulus of elasticity is calculated via the quotient of the difference in tensile stress and the difference in longitudinal strain by

$$E_f = \frac{\sigma_{f2} - \sigma_{f1}}{\varepsilon_{f2} - \varepsilon_{f1}}[MPa]$$

where: σf_2 is the stress measured at a strain value ε_2=0.0025 (0.25%); σ_{f1} is the stress measured at a strain value ε_1=0.0005 (0.05%).

The bending strength σ_f as resistance to bending load is calculated by

$$\sigma_f = \frac{3 * F_{max} * L}{2 * b * h^2}[MPa]$$

where: F_{max} is the maximum bending force [N]; L is the span [mm]; b is the width of the specimen [mm]; h is the thickness of the specimen [mm].

3 Results

The use of a pressure pot has the advantage of being able to process small quantities of a variety of systems without major cleaning effort. A disadvantage is the preparation quantity, which leads to a shortening of the pot life due to the exothermic effect. The batch was preheated to 60 °C. This temperature was derived from the viscosity values determined, see Fig. 4, and the modelled reaction kinetics. At 60 °C the viscosity is low enough allowing a working window of ~10 min to fill the mould with resin. This time dependent viscosity also reveals that the system GP56 + LV15877 (LV15877 and LV15888 are newly developed biosourced fatty acid poly aminoamides hardeners) has a lower viscosity compared to GP56 + LV15888. As later seen the thermo-mechanical properties of the two systems are more or less equal, therefore providing the hardener LV15877 is an advantage for processing the NFC.

Depending on the resin-hardener mixtures investigated, the conditions were thus different. By adjusting the injection pressure and finally also the mould temperature, the filling capacity of selected systems was investigated. Figure 5 shows the influence of two different hardness viscosities: LV15877 with 2720 mPa * s and LV 15888 with 7200 mPa * s. While filling at approx. 8 bar is possible with the LV 15877 system, even an increased pressure of approx. 9.5 bar is not sufficient for filling the cavity with the LV 15888 system. If the mould temperature is also increased, the unfilled volume increases further. In contrast, filling with LV 15877 is possible even at 115 °C mould temperature. When the maximum of the laboratory plant was reached with approx. 10 bar and 115 °C, the tests were successfully continued on an industrial plant: The complete filling of a plate cavity of 500 × 500 × 2 mm was possible without any problems.

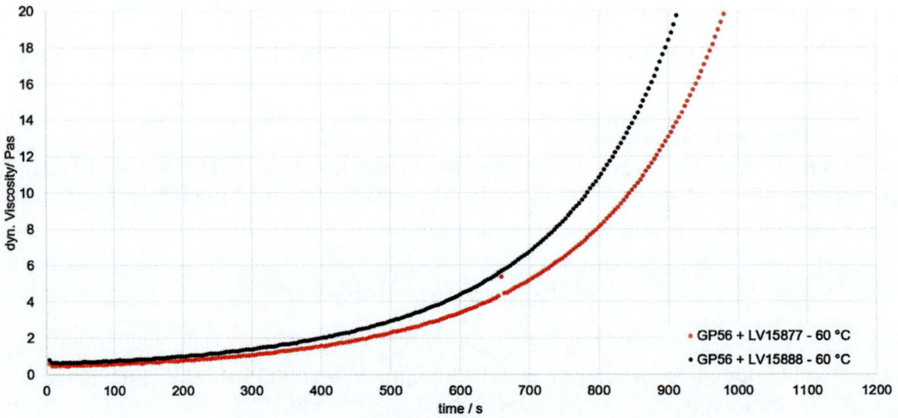

Fig. 4. Time-dependent viscosity of GP56 + LV15877 and GP56 + LV15888

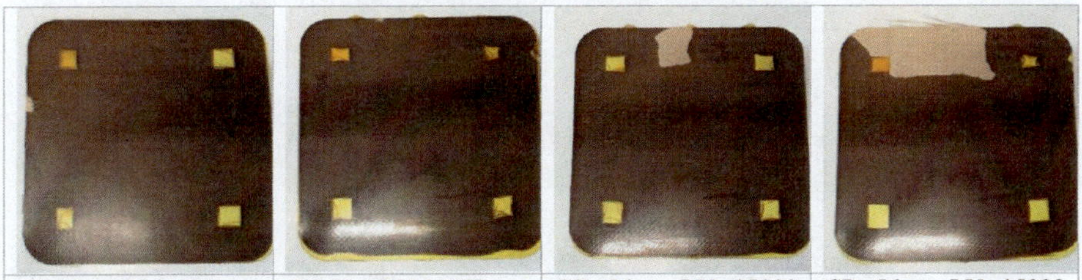

GP 56 + LV 15877 Mouldtemp.: 105°C Pressure: 8 bar	GP 56 + LV 15877 Mouldtemp. 115°C Pressure: 8 bar	GP 56 + LV 15888 Mouldtemp.: 105°C Pressure: 9,5 bar	GP 56 + LV 15888 Mouldtemp. 115°C Pressure: 9,5 bar

Fig. 5. Filling level of different hardeners with different process parameters

The results obtained in this work, show that to maximize the conversion α and thus the optimum thermal-mechanical properties, it is necessary to increase the process temperature and/or extend the processing time. Figure 6 shows the relative energy consumption of different variations of time and temperature. The energy consumption shown, includes the temperature control of the mould, the vacuum, as well as the demand of the press and the injection pressure. As an example, the parameters for the mixture GP 56 + SZ 8525 specified in material and methods were determined as 100% in Fig. 6. A relative representation was chosen, as the energy consumption is tool- or system-specific and absolute values are not necessarily representative for the targeted component. The results illustrate that an increase in temperature by 10 does not lead to a significant increase in energy consumption. If, on the other hand, the temperature is increased by 20 K, a reduction in time is necessary to prevent an increase in energy demand. At a constant mould temperature, an increase of curing time leads to a significant increase in energy demand of approx. 50%. The consideration of the target value glass transition temperature by means of DMA shows on the one hand that this longer time does not increase the glass transition temperature. Only an increase in the mould temperature leads to an increase in the glass transition temperature. On the other hand, it shows that the selected reference is not fully implemented. However, it should be noted that these investigations will be necessary as soon as a target-oriented resin-hardener formulation is found. This is to be compared with the petroleum-based reference in order to define extended process conditions on the basis of energetic variables.

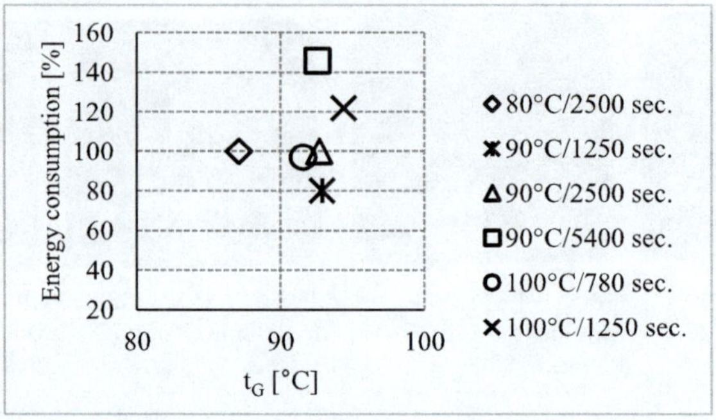

Fig. 6. Relative energy demand and glass transition temperatures as a function of temperature and time

The stiffnesses and strengths of the composite are decisive in determining whether the target component can be reshaped without adapting the layer structure. Figure 7 (right) shows the tensile strength of Greenpoxy 56 with selected developed formulations. The mean values of the tensile strengths compared to the petroleum-based reference have similar values compared to the reference system. This is also the case with the young modulus Fig. 7 (left).

Other important characteristics are the flexural modulus and the flexural strength of the NFC, Fig. 8, left shows the flexural moduli of the NFCs. For the bio sourced NFC similar values compared to the petroleum-based reference were found. The same holds for the flexural strength (Fig. 8, right).

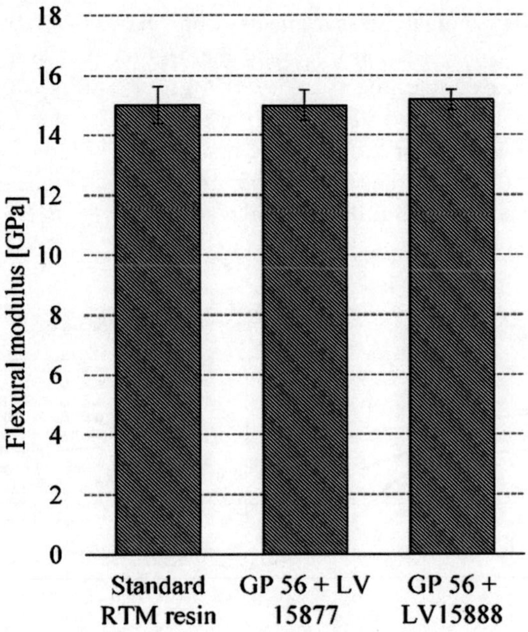

Fig. 7. Young's modulus (left) and tensile strength (right) of different formulations

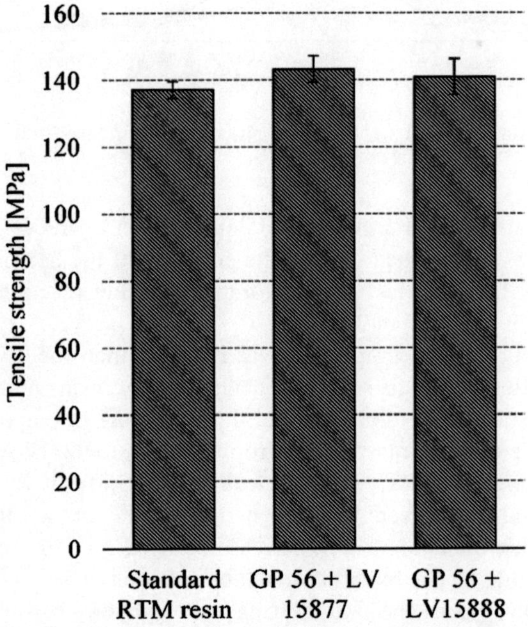

Fig. 8. Flexural modulus (left) and flexural strength (right) of different formulations

Further thermodynamical measurements were carried out by means of DMA. Figure 9 shows the Young's moduli, E', and the loss factor, tan δ, of fibre-reinforced epoxy resins. As an example, the system GP56+LV15888 is depicted. The samples were cured at either 110 °C or 150 °C for 20 or 90 min. The sheer epoxy resin without fibres was cured at 150 °C for 20 min. The results clearly indicate that the flax fibre has an influence on the glass transition temperature of the composite. A curing temperature of 150 °C for 20 min will fully cure the resin as proven by DSC-measurements.

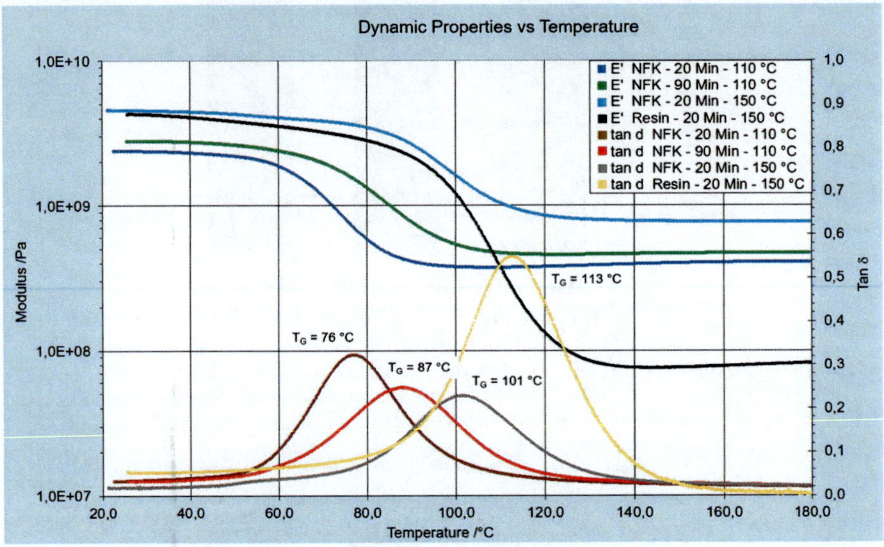

Fig. 9. DMA results of GP56+LV15888 NFC and native resin

Figure 10 reveals the isothermal conversion at 127 °C, directly measured by determining the conversion via integration of the enthalpy of the heat flux and by calculation via the model-free kinetics. After 20 min the curing reaction is almost finished with a conversion of higher than 93%.

However, the T_G of the composite is 12 K lower than the T_G of the resin without fibres. This difference is also found for other epoxy resins under investigation. To determine the reason for this temperature, drop the cured composite was stored for 48 h in water. After water storage the T_G dropped down to 92 °C. After drying at room temperature for 2 weeks, the T_G went back to 98 °C. So far it can be concluded that water and moisture have a reversible influence on the T_G on the fully cured composite in terms of softening the composite. The reason why the T_G of the composite is 12 K lower than that of the sheer resin can only be speculated on. It is possible that the moisture in the fibre opens the oxirane rings of the epoxy resin which are then not available to cross-link with the hardener. The non-cross-linked fatty acid chains may soften the cured epoxy resin. Further investigation in order to prove this hypothesis are ongoing.

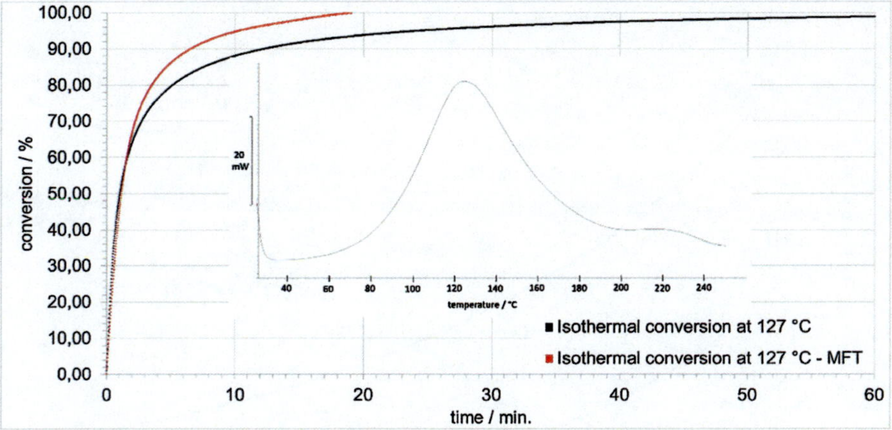

Fig. 10. Isothermal conversion of GP56+LV15888 at 127 °C; DSC heat flux

To estimate the required curing parameter of the resin the conversion of resin was measured by DSC. The isothermal conversion at 127 °C reveals that approximately 20 min are required to achieve.

95% curing rate (Fig. 11). The applied model-free kinetic technic does not work for this system. This is because of a reaction that has its maximum value at 220 °C. As the temperature cannot exceed 250 °C in the DSC because degradation processes of the resin start, the end of the reaction cannot be fully monitored. Hence, the identity of the isothermal conversion and the model-free kinetic fits very good up to 65% conversion, but differ with higher conversion rates.

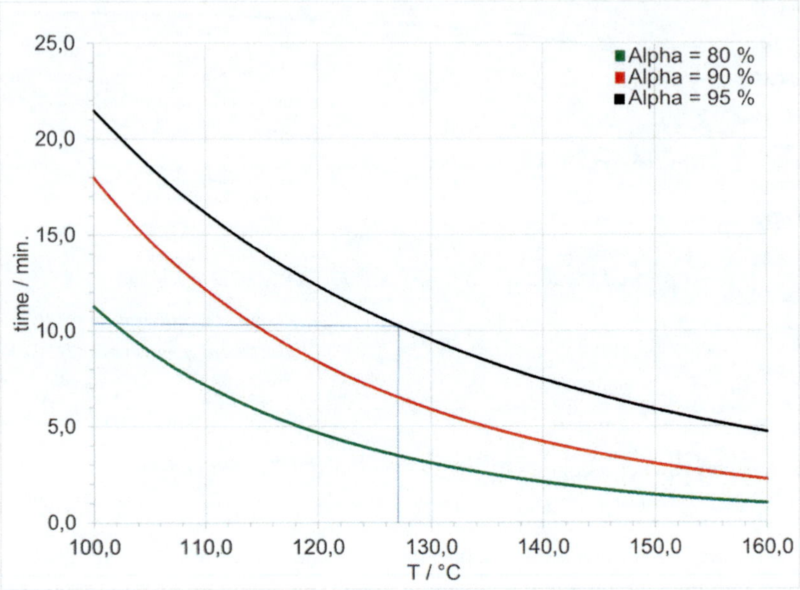

Fig. 11. Isothermal conversion of GP56+LV15888 at 127 °C

4 Conclusion and Outlook

Until now, only a partial bio-sourced epoxy resin was available. Within this project we managed to develop also partial bio-sourced hardener based on fatty acid aminoamide technology. The composites have therefore a biosourced carbon content of >75%. The aim to generate a T_G higher than 100 °C for the NFC could be achieved. The developed system only needs a temperature above 125 °C for 20 min to accomplish an appropriate curing. This is 35 K more than conventional bis-phenol-A-type petrol-sourced epoxy systems, that requires only 90 °C. However, first calculation exposes, that a higher temperature is only slightly worsening the energy balance of the production procedure. The curing time is much more critical. The curing time of the biosourced epoxy is, however, comparable to the conventional epoxy system.

One point to be focused on in future work is the water influence on the curing and maximum achievable T_G. Ongoing work shall allow a better understanding of the influence on the side reaction of the epoxy and the reversible influence on the T_G. Pre-treatment of the fibres, e.g. by impregnation, may diminish these effects.

Acknowledgements. The authors would like to thank for the financial support by the Federal Ministry of Food and Agriculture (BMEL), through the FNR (Fachagentur Nachwachsende Rohstoffe e. V.) based on a decision taken by the German Bundestag. Funding no. 2220NR094A and 2220NR094B.

References

1. Le Duigou, A., Davies, P., Baley, C.: Environmental impact analysis of the production of flax fibres to be used as composite material reinforcement. J. Biobased Mat. Bioenergy **5**(1), 153–165 (2011). https://doi.org/10.1166/jbmb.2011.1116
2. Hedlund-Åström, A.: Model for end of life treatment of polymer composite Materials. Doctoral thesis, Royal Institute of Technology. https://www.diva-portal.org/smash/get/diva2:12611/FULLTEXT01.pdf (2005). Accessed 20 Apr 2022
3. Stiller, H.: Material intensity of advanced composite materials. Results of a study for the Verbundwerkstofflabor Bremen e. V. Wuppertal Papers, 90. Wuppertal Institut für Klima, Umwelt, Energie. https://epub.wupperinst.org/frontdoor/deliver/index/docId/926/file/WP90.pdf (1999). Accessed 20 Apr 2022
4. Netravali, A.N., Chabba, S.: Composites get greener. Mater. Today **6**(4), 22–29 (2003). https://doi.org/10.1016/S1369-7021(03)00427-9
5. Mittal, V., Saini, R., Sinha, S.: Natural fiber-mediated epoxy composites – A review. Compos. B Eng. **99**(7), 425–435 (2016). https://doi.org/10.1016/j.compositesb.2016.06.051
6. Vigneshwaran, S., et al.: Recent advancement in the natural fiber polymer composites: A comprehensive review. J. Clean. Prod. **277**(6/7), 124109 (2020). https://doi.org/10.1016/j.jclepro.2020.124109
7. Nassar, M., Arunachalam, R., Alzebdeh, K.: Machinability of natural fiber reinforced composites: a review. Int. J. Adv. Manuf. Technol. **88**(9–12), 2985–3004 (2016). https://doi.org/10.1007/s00170-016-9010-9
8. Torres-Arellano, M., Renteria-Rodríguez, V., Franco-Urquiza, E.: Mechanical properties of natural-fiber-reinforced biobased epoxy resins manufactured by Resin Infusion Process. Polymers **12**(12) (2020). https://doi.org/10.3390/polym12122841
9. Petrie, E.M.: Epoxy adhesive formulations. McGraw-Hill chemical engineering. McGraw-Hill, New York (2006 i.e. 2005)
10. Baroncini, E.A., Kumar Yadav, S., Palmese, G.R., Stanzione, J.F.: Recent advances in bio-based epoxy resins and bio-based epoxy curing agents. J. Appl. Polym. Sci. **133**(45), 842 (2016). https://doi.org/10.1002/app.44103
11. Ramon, E., Sguazzo, C., Moreira, P.: A review of recent research on bio-based epoxy systems for engineering applications and potentialities in the aviation sector. Aerospace **5**(4), 110 (2018). https://doi.org/10.3390/aerospace5040110
12. Mustapha, R., Rahmat, A.R., Abdul Majid, R., Mustapha, S.N.H.: Vegetable oil-based epoxy resins and their composites with bio-based hardener: a short review. Polymer-Plastics Technol. Mater. **58**(12), 1311–1326 (2019). https://doi.org/10.1080/25740881.2018.1563119
13. François, C., et al.: Design and synthesis of biobased epoxy thermosets from biorenewable resources. C. R. Chim. **20**(11–12), 1006–1016 (2017). https://doi.org/10.1016/j.crci.2017.10.005
14. Paramarta, A., Webster, D.C.: Bio-based high performance epoxy-anhydride thermosets for structural composites: The effect of composition variables. React. Funct. Polym. **105**, 140–149 (2016). https://doi.org/10.1016/j.reactfunctpolym.2016.06.008
15. Merighi, S., Mazzocchetti, L., Benelli, T., Giorgini, L.: Adenine as epoxy resin hardener for sustainable composites production with recycled carbon fibers and cellulosic fibers. Polymers **12**(12) (2020). https://doi.org/10.3390/polym12123054
16. Zhang, K., Gu, Y., li, M., Zhang, Z.: Effect of rapid curing process on the properties of carbon fiber/epoxy composite fabricated using vacuum assisted resin infusion molding.

Materials & Design (1980–2015) **54**(2), 624–631 (2014). https://doi.org/10.1016/j.matdes.2013.08.065
17. Kanny, K., Mohan, T.P.: Resin infusion analysis of nanoclay filled glass fiber laminates. Compos. B Eng. **58**(2), 328–334 (2014). https://doi.org/10.1016/j.compositesb.2013.10.025
18. Govignon, Q., Bickerton, S., Kelly, P.A.: Simulation of the reinforcement compaction and resin flow during the complete resin infusion process. Compos. A Appl. Sci. Manuf. **41**(1), 45–57 (2010). https://doi.org/10.1016/j.compositesa.2009.07.007
19. Vyazovkin, S., Wight, C.A.: Model-free and model-fitting approaches to kinetic analysis of isothermal and nonisothermal data. Thermochim. Acta **340–341**, 53–68 (1999). https://doi.org/10.1016/S0040-6031(99)00253-1

Investigation on the Bond Performance in Hybrid Wood-Plastic Components

Vicky Reichel(✉), Werner Berlin, Tim Ossowski, Yvonne Phung, and Klaus Dröder

Institute of Machine Tools and Production Technology, Technische Universität Braunschweig, Braunschweig, Germany
{v.reichel,w.berlin,t.ossowski,yvonne.phung, k.droeder}@tu-braunschweig.de

Abstract. Sustainable resource and energy management are essential aspects for the change of industrial products and processes of the coming years. The demand for resource-saving and sustainable products is increasing in many production areas. As a renewable resource with a wide range of properties, bio-based materials, such as wood, play a key role in this context. For the development of sustainable, lightweight as well as load-bearing structural components, the use of wood veneer shows great potential. Due to the microscopic structure of wood as cellulose fibers embedded in a lignin matrix, it is comparable to the structure of fiber-reinforced plastic semi-finished products such as tapes or organo sheets. In order to be able to serve a similar field of application as plastic-based semi-finished products, however, the wood-based product must meet the requirements in terms of functionalization, mechanical properties and process integration during production. One way to meet the specific conditions is to combine layered wood veneers together with plastic applications by means of wood-plastic hybrid components. The resulting plywood serves the function of bearing the load, while the plastic structure, e.g. ribs, ensures a sufficient stiffness of the part. Its structural integrity is mainly determined by the bond strength of the wood-plastic interface. The bond strength in turn depends on the type of plastic used, on the conditioning of the plywood and on the process parameters during overmoulding. In the context of this work, the bond strength between beech wood-veneers and different plastics polypropylene is investigated. Therefore, rib specimens are manufactured in an overmoulding process with varying parameters. Afterwards, the bond strength is determined in a rib pull-off test and the influence of the variation parameters is investigated.

Keywords: Hybrid wood-plastic components · Injection moulding · Bond strength

K. Dröder and T. Vietor (Eds.): *Future Automotive Production Conference 2022*, Zukunftstechnologien für den multifunktionalen Leichtbau, pp. 127–143, 2023.
https://doi.org/10.1007/978-3-658-39928-3_10

1 Introduction

The use of flat, fibre-reinforced semi-finished plastic products such as organo sheets is common in the automotive industry [1, 2]. They can be used monolithically [1] or in combination with other materials and production processes as hybrid components [2] as a coupling layer [3] or a load-bearing or reinforcing element [4]. However, the recyclability as well as the energy input during production and the environmental compatibility in general are low compared to bio-based alternatives [5]. Especially in the context of a stricter legal framework for sustainable mobility and a growing understanding of environmental responsibility among automotive manufacturers, the substitution of plastics by bio-based alternatives is targeted [6, 7]. The microscopic structure of wood is comparable to the structure of fibre-reinforced semi-finished plastic products: wood fibres are embedded in a matrix of lignin and hemicellulose [8]. Wood in general in structural parts is mainly used to take static loads [9]. For structural applications with dynamic loads, solid wood in form of veneer ply structures are currently mainly used monolithically [10]. The disadvantage is that sufficient stiffness of the ply structure can only be achieved by complex forming processes. Due to the limited formability of wood in general because of its high stiffness and very low ductility parallel to the grain direction [11], the focus is therefore on components that have low degrees of forming or are only bent around one axis. To compensate for this disadvantage, one possible approach is to combine it with other materials [12]. A suitable combination of materials is processing together with plastics. The combination of wood and plastic is mainly used in the form of wood-plastic composites (WPC). The wood raw material (flour, shavings, chips, fibres) are mixed into the polymer granules. Typical methods to produce wood-plastic-composites parts are direct extrusion, injection moulding or pressing techniques [13]. The wood component mainly serves as a filler or, to a limited extent, as a load-bearing material. [14]

In particular, primary forming manufacturing processes such as extrusion or injection molding offer a high degree of flexibility with regard to the part's geometry. So the plastic application can compensate for the limited formability of the wooden component and even flexibly form other complex geometries such as connection points to other parts without further processing. With regard to the stiffness of a hybrid component made of plastic and wood, a plastic rib structure can provide sufficient stiffness and the wood veneer or a veneer layer structure can absorb mechanical loads [12].

The use of wood veneers in combination with other material as hybrid components is already established in the field of decorative surfaces [15–18]. Thin veneer is processed in injection moulding and overmoulded as well as back-injected with ribs and other supporting structures. For load-bearing structural components, there are only initial approaches to combining plastics and wood veneers [19].

For the production of plastic-wood hybrid components it is necessary, to adapt existing plastic processing methods and to apply the processing parameters to the material restrictions of the two composite partners wood and plastics. In particular, the bond strength between the two partners is decisive for the mechanical properties of the overall component [20]. The bonding of plastics to wood has so far only been investigated in the form of full-surface contact between veneer and a plastic structure produced in an injection moulding process [21] or plastic-extrusion [22]. The local bonding of plastics on wood veneers has only been investigated initially [23].

The aim of this work is therefore to evaluate the influences of the manufacturing parameters of injection moulding and the conditioning of a wood veneer on the bond strength between wood veneer and overmoulded plastic ribs.

2 Bonding Mechanisms in Wood-Plastic Hybrid Components

In principle, wood and plastic can form chemical and mechanical connections, especially at the microscopic level in the boundary layer between the two bonding partners. Figure 1 shows the cross section of beech veneer on the microscopic scale. Here, a sequence of secondary xylem and medullary rays can be seen. The secondary xylem consists mainly of tracheae, which form thin capillaries with the cell walls for water and nutrient transport in the z-direction. The medullary rays extend transversely to these capillaries in the x-direction. They are responsible for water and nutrient transport in the radial direction as well as the mechanical properties of the wood. The medullary rays are composed of cells that form a continuous capillary.

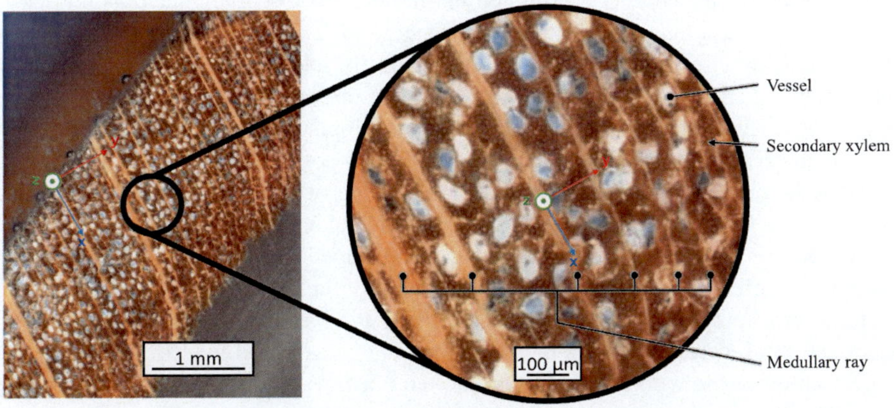

Fig. 1. Microscopic structure of beech veneer with its main constituents

The cell walls of wood cells consist mainly of cellulose, hemicellulose and lignin. Cellulose accounts for the highest proportion of wood-based materials, at approx. 45% by mass. [24] Cellulose is a polymer (polysaccharide) from the monomer cellobiose. Cellobiose, in turn, is a disaccharide from β-glucose [25]. The chemical structure of cellulose is shown in Fig. 2.

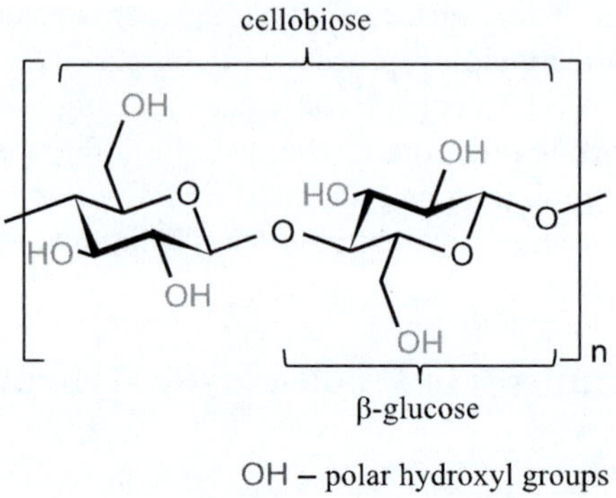

Fig. 2. Chemical structure of cellulose as a polymer from cellobiose

Cellulose has hydroxyl groups at its free ends. Due to the higher electronegativity of oxygen compared to hydrogen, these are partially electrically positively charged. Cellulose thus has the potential to form weak physical bonds via hydrogen bonds. This occurs on the one hand within the molecular chain, but also outwardly to other bonding partners. Hydroxyl groups are highly reactive and can also form chemical bonds. [26]

In addition to the possible chemical bonding of cellulose to other materials, wood can also realize mechanical bonding at the micro level, particularly in combination with plastics. By penetrating the molten plastic into microscopic undercuts on the surface of the wood, a micro-form-fit can be established. These interlocking joints provide resistance against separation and thus a mechanically relevant amount of bond strength between wood and plastic. A high penetration pressure at the materials interface as well as a low viscosity of the plastic melt favour the development of the bond strength. A possible chemical bond between wood and plastic is also positively influenced by a low viscosity of the plastic melt, as this enhances the wetting of the microscopic wood surface. Hence, more chemical joints can be established and the bond strength increases. [27]

In addition to the process parameters during injection molding, the viscosity of the plastic melt is also significantly determined by the temperature of the melt [28]. As long-term temperatures over 130 °C damage the basic components of wood, it is beneficial to use plastics with a melting point as low as possible [19]. To improve the impregnation of the wood surface with molten plastic, the zero shear viscosity should also be as low as possible, even at low temperatures close to the melting point.

Therefore, polypropylene is predominantly used in Germany for wood-plastic-compound (WPC) applications [13]. Polypropylene belongs to the group of polyolefins and is partially crystalline and non-polar. This means that there are no charge centres at the branches of the molecule chains. [29] The missing charges have an influence on the possible bonding mechanisms that polypropylene can establish to wood. Chemical or physical bonds to the hydroxyl groups of cellulose are not possible with pure polypropylene. In order to achieve a chemical bond, polarization is only possible by a grafting reaction in which polypropylene reacts radically with polar, unsaturated compounds. One of the most common applications is the use of maleic anhydride grafted polypropylene (MAHg-PP) as a bonding agent. [30]

Through an esterification, the hydroxyl groups of cellulose react with the grafted maleic anhydride molecule and form a chemical bond [31]. An increased temperature supports the formation of the chemical bond between the aromatic ring of the malic anhydride and the hydroxyl groups of the cellulose [32]. Accordingly, the temperature control in the injection molding process is decisive for the strength of the wood-plastic hybrid components.

3 Methods

3.1 Specimen Geometry and Mould

For the determination of the bond performance between wood and plastic, suitable specimens were designed and manufactured in an injection moulding process (Fig. 3). The fibre orientation of the wood veneer is transverse to the longitudinal axis of the ribs and is oriented in the z-direction due to better testing characteristics (see Fig. 3 and Sect. 3.3). The base geometry of the ribs influences the notch effect when transferring forces from the rib to the veneer. The rib with the wide base transfers the forces more homogeneously, which is why it can be assumed that higher bond strengths will be achieved. Also the increased base area leads to more surface available for chemical and mechanical bond.

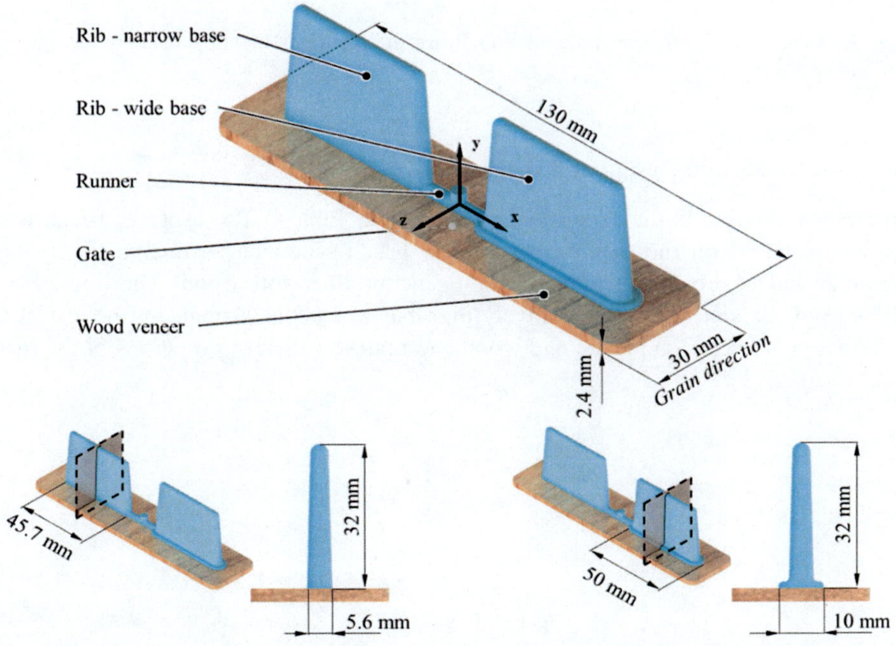

Fig. 3. Rib pull-off test specimen as injection moulded component with veneer insert consisting of narrow and wide rib base

The moulding device for producing the rib pull-off test specimen is part of an injection moulding mould that provides slide-in units for a quick mould change (Fig. 4). The cavity is only located in the movable mould half. The fixed mould half consists of a flat, plain counter plate which only contains the sprue. The fixed mould half has no further effect on the moulding process. Fluid (water) by means of the temperature control connectors temper the mould unit. After the component has been cooled to demoulding temperature, it was ejected via integrated ejector pins.

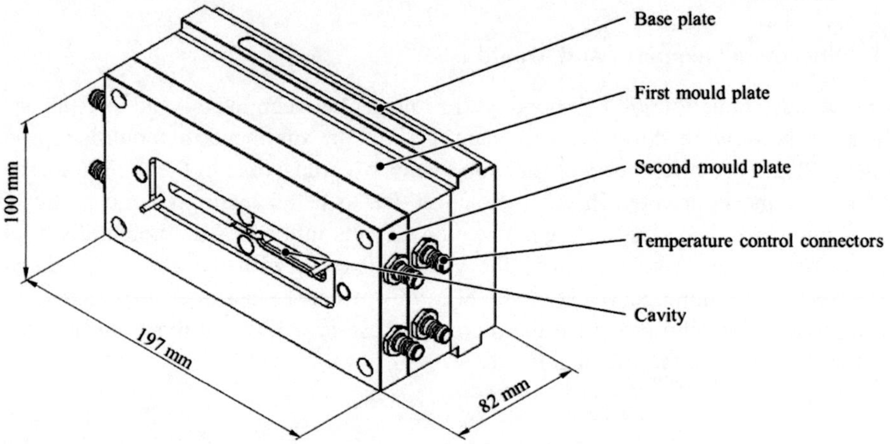

Fig. 4. Mould unit for the movable mould half for the specimen manufacturing in the injection moulding process

3.2 Materials and Design of Experiments

The materials used in the experiments are listed in Table 1. The usage of beech wood veneer is based on the mechanical values: beech wood has a comparatively high strength and is grown much more homogeneous than soft wood. The thickness of 2.3 mm of the veneer is comparatively high but was needed to ensure the test of the adhesive bond between plastic and wood and not the cohesive properties of the materials itself.

Table 1. Materials and Parameters

Used Materials	
Wood	Beech veneer, thickness: 2.3 mm
Plastic	lyondellbasell—Moplen HP501H, dried before use
Bonding Agent	ExxonMobile® Exxelor™ PO 1020 (containing approximately 1 wt.-% of maleic acid [33])
Variation Parameters	
Wood moisture content u (WMC)	u = 7%, 12%
Mould temperature T_m	50 °C; 70 °C
Injection temperature T_i	230 °C, 250 °C
Rib geometry	Narrow (A: 45.7 mm × 5.6 mm), Wide (A: 50.0 mm × 10 mm)

Table 1 shows as well the process and material parameters varied in this experimental study. As mentioned before, there is a significant influence on the bond strength expected by changing the wood moisture content. Due to the relation of wood moisture to the specimen geometry and wood structure [34], the moisture content was evaluated before the manufacturing of the specimen Usually the wood moisture content (WMC) of beech wood materials at standard climate (23 °C / 65% rel. h.) is u ~ 12% [35]. The laboratory climate used was 22 °C/50% rel. h. which leads to u ~ 7% at the selected specimen geometry. Due to comparability the WMC of u = 12% was also set in the experiments. The time needed to condition the veneer specimen and the wood moisture reached at two different levels of air humidity is shown in Fig. 5.

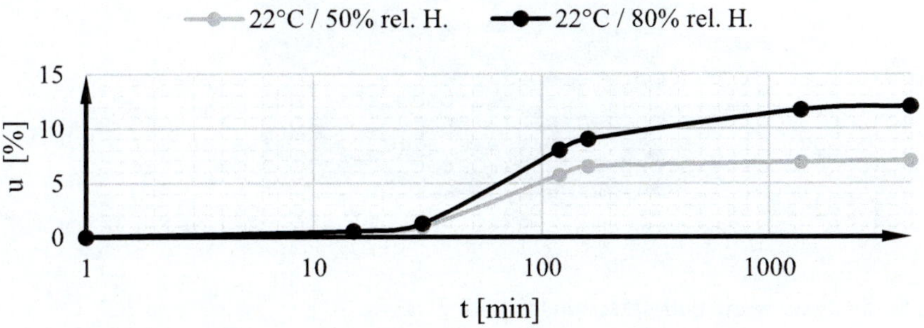

Fig. 5. Specific wood moisture content for two different air humidity levels

By changing the mould temperature T_m and injection temperature T_i it was expected to extend the period of critical temperature level needed for the chemical bonding. Furthermore, the viscosity is kept at a lower level for a longer period of time, which allows better penetration of microstructures on the wood surface. The variation of the rib geometry allows to describe different fail modes of the veneer in the bonding area. It was expected to reach higher values for the pull-off force with the wide base geometry due to a comparatively favorable load path design.

Since wood is a naturally growing material, the structure and therefore the mechanical characteristics can vary in high amounts. Therefore, a full parameter DOE was applied to reach the highest possible validity. The parameters were changed on two parameter levels to describe linear relations to the target value of bond strength. There were 80 specimen manufactured and tested. The data analysis follows a double sided significance test with a p-value of 0.05.

3.3 Test Method

The bond strength can be determined by applying of a tensile force onto the interface between the bonding materials of the hybrid specimen. The test method used to evaluate the bond strength follows the specimen structure used by e.g. [36 and 37] for hybrid material combinations. There is neither a standard method for the test nor the specimen.

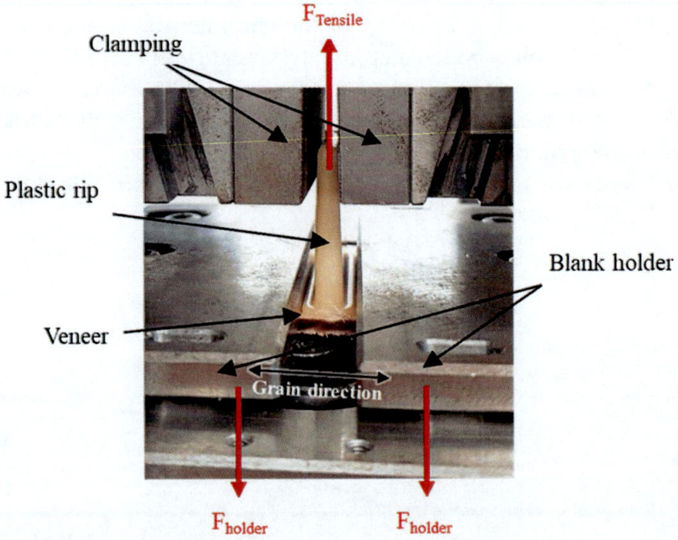

Fig. 6. Setup of the rib pull-off test in the testing machine

Due to the low thickness of the veneer, it is necessary to apply the clamping force via a blank holder taking into account the growing direction of the wood (see Fig. 6). The fiber orientation should be 90° angled to the rib and blank holder edge. Using lower angles or 0°, it is possible to test the bending strength of the veneer itself. The testing speed was v = 5 mm/min which leads to bond fail in t = 60 s ± 15 s.

4 Interpretation and Conclusion

The determined bond strengths were between 40 and 1500 N. One significant point to mention was the specimen quality. There were 24 of 80 specimen with edge flashing as it can be seen exemplarily in Fig. 7. The regression model was created with all forces measured, including the edge flashes.

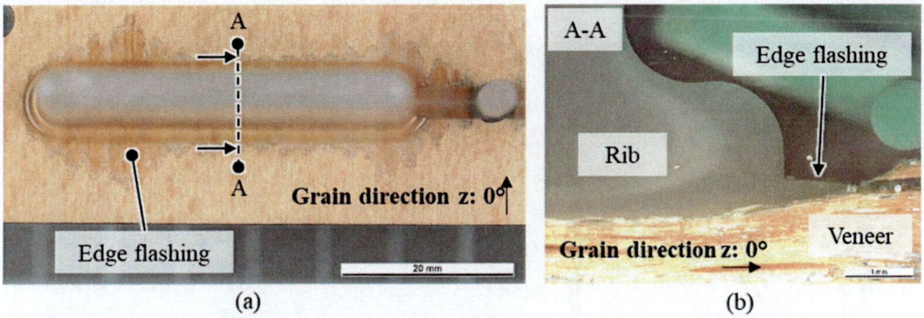

Fig. 7. Edge flashing (a) top view, cut plane; (b) micro-cross section view (A-A)

The edge flashing phenomenon occurs more frequently when the wood moisture content (WMC) was increased and the temperatures of T_m and T_i were higher. When the edge flashings occur the bonding area increases unspecifically and by means of the setup it was not possible to clamp the specimen only at the veneer structure. Therefore, the determined forces were increased but referred to the rib base surface. The effect of edge flashings were detected at all specimen with a resulting pull-off force over 800 N(wide base) and 440 N(narrow base). It was not possible to influence the edge flashing effect positively to get higher quality specimen at the specified parameters.

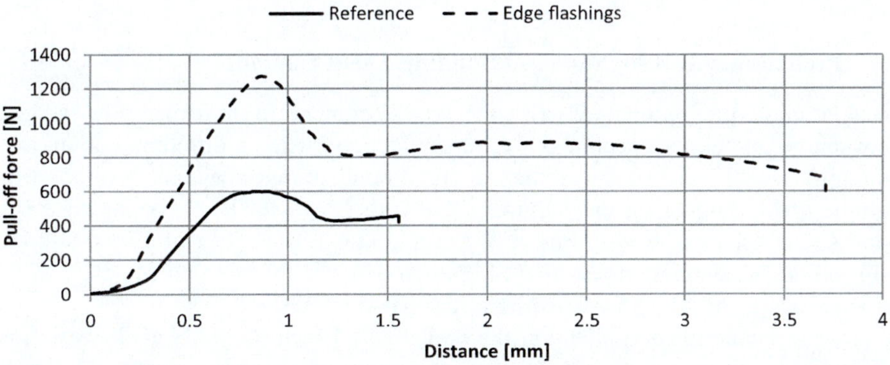

Fig. 8. Exemplary graphs of a reference (full) and specimen with edge flashings (dashed)

In Fig. 8 are two graphs of a reference and a specimen with edge flashings shown. While the force for edge flashing is higher, the failure point of this specimen is also extended up to 3,7 mm. This effect occurs due to the different failure behavior. At Fig. 9 there is an exemplary specimen with edge flashes showed. When looking at the bonding surface sticking wood fibers can be seen, on base area but as well at the edge flashings. The load at the area of edge flashes is not brought vertical into the bond area. The area of edge flashes is loaded more horizontally which is comparable to shear stresses.

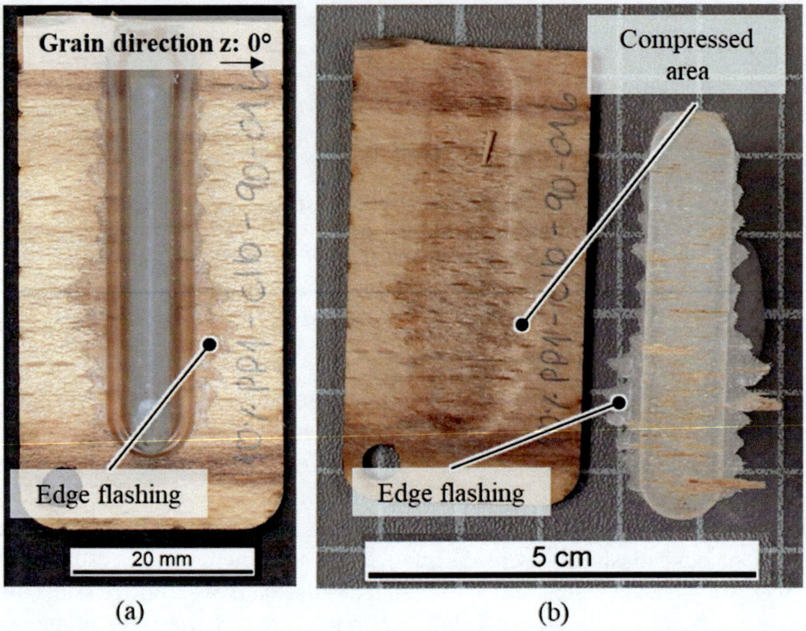

Fig. 9. Exemplary specimen (a) untested: edge flashings; (b) tested: compressed area and sticking wood fibers on plastic

4.1 Preliminary Investigations on Bonding Agent Content

To determine the optimal MAH-g-PP wt.-% content to maximize the adhesion between the materials, specimens were manufactured using a different amount of the bonding agent. Figure 10 shows the measured bond strength depending on the filling volume of MAH-g-PP in the Polymer from 0 to 15 wt.-%. It can be stated that the bond strength increases with higher MAH-g content. The effect of 15 wt.-% MAH-g-PP on the bond strength was +0.72 MPa compared to the strength achieved with 10 wt.-% MAH-g-PP, which was also significant (p-val.:0.03).

Due to economical application the content of 10 wt.-% MAH-g-PP was chosen (effect: +0.64, p-val.: 0.28) for further investigations. Higher contents of bonding agents can affect the properties of the PP raw material and result in poorer environmental compatibility.

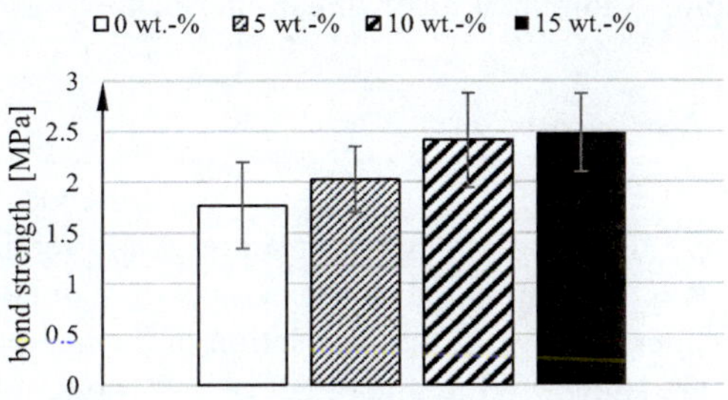

Variation
MAH-g-PP wt.-%
Mould temperature
50 °C
Injection temperature
230 °C
Rib base geometry
narrow
Wood moisture content (WMC)
u = 7 %
Testing parameter
v_t: 1 mm/min; n: 5; 22°C/50% rel. H.

Fig. 10. Results on the variation of MAH-g-PP wt.-% content in PP material and its influence on the bond strength

By analysing the results, it is also possible to estimate the strength of the mechanical form-fit. Comparing the two mechanisms it can be stated that the mechanical effect is higher (1,77 MPa) than the chemical bond (at 15 wt.-% MAH-g-PP: 0.72 MPa) (Fig. 11).

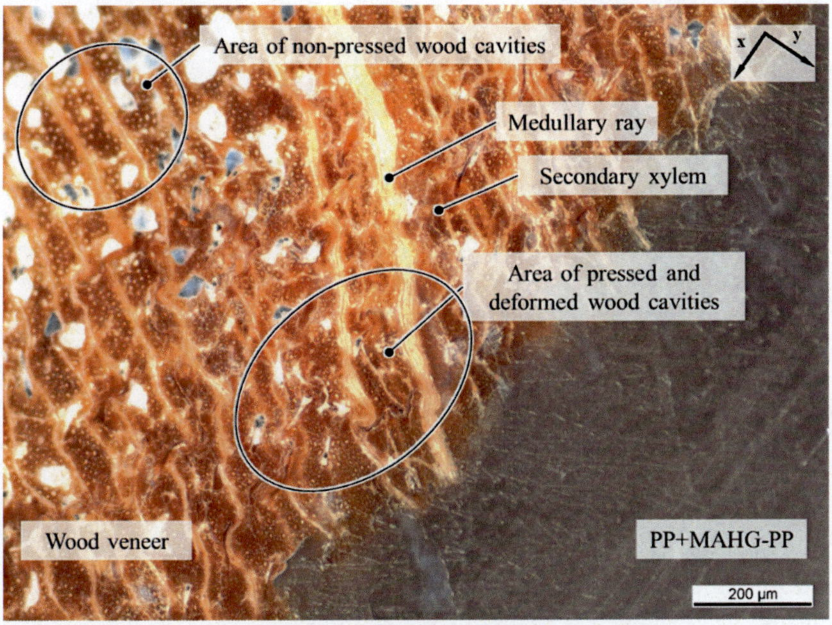

Fig. 11. Micro cross-section at the material interface of wood and plastic, deformation of pressed wood veneer in x/y axis

The plastic melt pressure deforms the veneer surface. The pores are closed so that there is no plastic interlock in this natural cavities possible. Also, the soft secondary xylem is highly deformed while the higher strength of the medullary rays leads to macroscopic interface increase. Therefore, a dominant mechanical interlocking mechanism is possible.

4.2 Results on the Bond Strength

The experimental determined values were used as data input for a regression model. In the course of this investigation it was determined that the temperature related parameters were the most influencing values. In Fig. 12 the correlation of the parameters onto the target value bond strength are shown. While the effect of the wood moisture content is very low the geometry of the rib even has an even lower and negative effect. This leads to the assumption that WMC and rib geometry do not influence the value of the bond strength significantly.

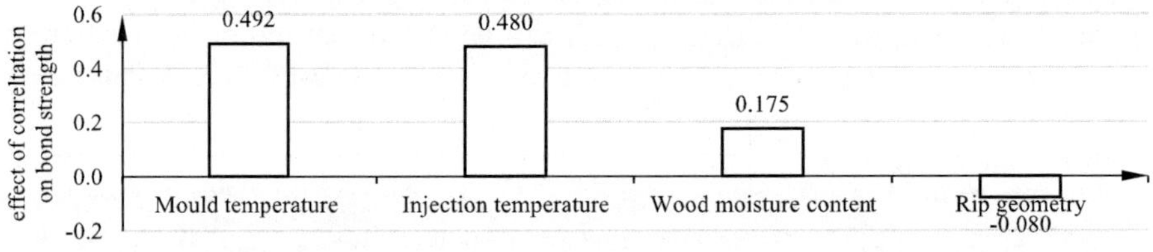

Fig. 12. Correlation of varied parameters to target parameter "bond strength"

Overall there were only slight interactions between these parameters detected. The highest interaction of varied parameters was detected between the values mould temperature and injection temperature (0.477). Therefore, an increasing temperature at both parameters influences the bond strength more positively than the variation of just one of the parameters. The higher process temperatures lead to extended time period of low viscosity of the melt and therefore to a better situation for the chemical bonding. The low viscosity level is also favorable to support injection pressures resulting as mechanical interlock effects at deformed wood surface.

The validity of the regression model and therefore its coefficients is at 0.8 which allows the assumption of 20% inexplicable deviation and only partial description of the connection between the parameters. The model is therefore not suitable to find optima parameter values.

Mould Temperature and Injection Temperature

The higher bond strength at higher mould and injection temperatures can be attributed to the bond promoting interface conditions (interlocking effects and chemichal bonding) as well as the enlargement of the connection area due to the edge flashing. The temperatures have therefore the highest effects on the bond strength T_m coefficient 0.598 (p-val.: 0) T_i coefficient 0.593 (p-val.: 0) and are highly significant as well.

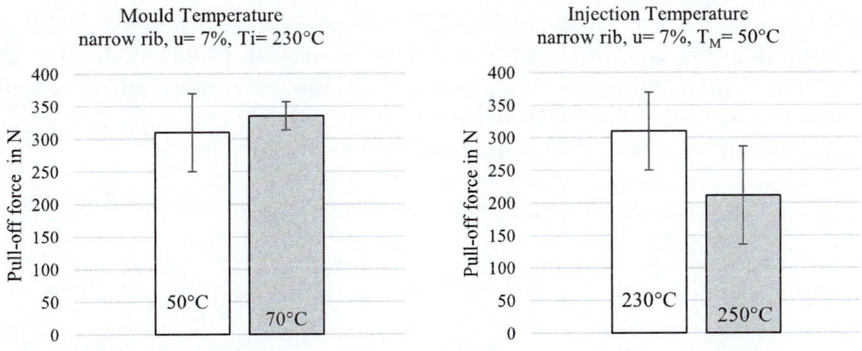

Fig. 13. Exemplary results for the variation of mould temperature and injection temperature

It is assumed that higher temperatures lead to better binding due to higher viscosity of the melt. Therefore, the surface microstructure is filled much better. Not all results are showing this connection (see Fig. 13). This can be justified by the low viscosity change of the melt, as the difference between the selected temperatures is only 20 °C.

Rib Base Geometry

Because it is already known that the rib base geometry has an influence on the bearable pull-off force between the materials, this parameter is varied in wide and narrow. Therefore, different load paths were represented. The results of this variation in correlation to the WMC are shown in Fig. 14.

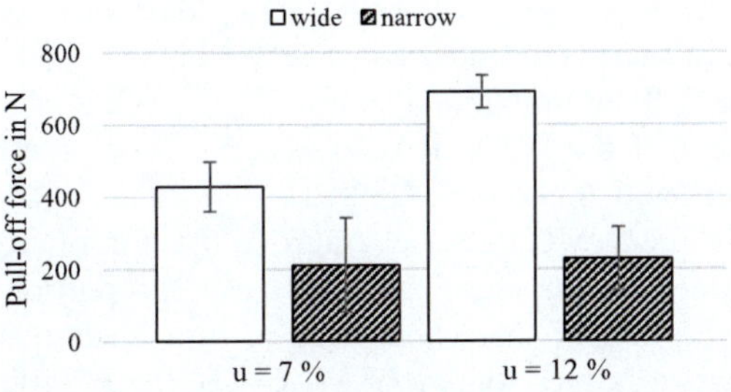

Variation
Climate conditions
Mould temperature
50 °C
Injection temperature
250 °C
Rib geometry
narrow, wide
Wood moisture content (WMC)
u = 7 %, 12 %
Testing parameter
v_t: 5 mm/min; n: 5; 22°C/50% rel. H.

Fig. 14. Pull-off forces for different base geometries

In 50% of the variation cases the narrow geometry shows higher pull-off forces. This leads to the assumption that the rib geometry has no significant influence. The analysis on interactions shows that the narrow geometry results in higher forces when the wood moisture content is lower at u = 7%. The test on the significance showed that based on the values in this research the rib geometry has no influence (coefficient: -0.12; p-value: 0.14).

Wood Moisture

The wood moisture content is one of the most influencing parameters on the wood characteristics [34]. When increasing the moisture in wood materials their strength is lowered. In case of this study the decreased compressive strength leads to edge flashing effects at higher WMC. Exemplary results can be found in Fig. 15.

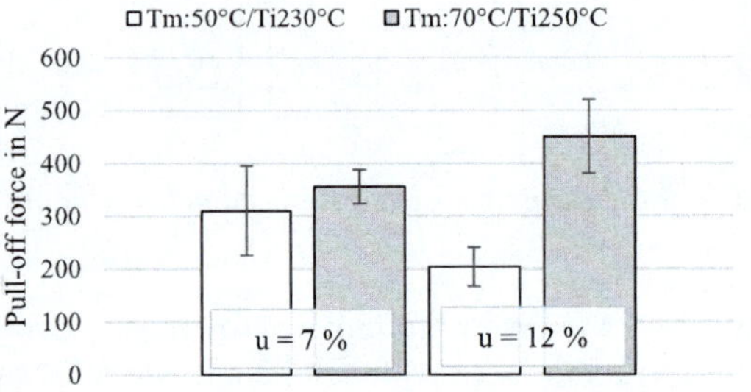

Variation
Climate conditions
Mould temperature
50 °C/70 °C
Injection temperature
230 °C/250 °C
Rib geometry
narrow
Wood moisture content (WMC)
u = 7 %, 12 %
Testing parameter
v_t: 5 mm/min; n: 5; 22°C/50% rel. H.

Fig. 15. Influence of low and high process temperatures at different WMC levels on the resulting maximum tensile force

A higher moisture content leads to better deformability of the wood surface. Therefore, it was expected that a better mechanical interlocking effect can be observed due to a more intense penetration of the melt into the surface of the wood. The presence of a higher amount of water molecules should also lead to a better chemical bonding. Both expectations could not be approved. The higher WMC leads to higher deformation but due to the high pressure of the melt, edge flashings occur permantly. Therefore, the bonding area increased undefinable and a high amount of the specimen tested were invalid (especially at mould temperature T_m: 70 °C, injection temperature T_i: 250 °C). The coefficient of the WMC is 0.26 which is, compared to the other parameters, low but significant with a p-value of 0.002.

It can be stated that an increase in the WMC does not improve the bond performance very much. This can be explained by the fact that an increased proportion of bonding agent inhibits the water absorption of the wood. This in turn leads to the assumption that a higher WMC requires a higher content of MAG-g in the PP granulate to achieve the same level of strength.

4.3 Conclusion

The investigations showed that the WMC, the mould temperatures and injection temperatures have an influence on the bond strength between wood and plastic. Further, higher pull-off forces can be reached with a narrow base geometry instead of a wide base geometry. A higher WMC leads to higher deformability of the wood veneer which in turn leads to edge flashings for 35% of the manufactured specimen. The edge flashing effect appears especially in combination with higher process temperatures (+59%). The highest value at 812 N/1.6 MPa without edge flashing was reached at narrow base geometry, u: 12%, T_m: 50 °C and T_i: 250 °C.

Overall, it can be stated that the reached bond strengths in the addressed hybrid wood-plastic component is low compared to common hybrid connections like over-moulded organo sheets or pretreated sheet metal. To be competitive in structural applications the wood-plastic connection needs to be improved. First steps to realize this should be the test of other plastic materials. PA6 for example as a polar raw material can be a suitable start for investigations but needs to be processed in short time cycles to avoid wood material damage by high temperatures. As second step the plastic material can be customized via additives as well as the wood material can be pretreated to improve connectivity of the materials interface.

Acknowledgements. This study was carried out within the project WPPro- "One-shot injection molding process for wood-plastic components" (KK5052901LL0, 2019–2022) which was funded by the Central Innovation Programme for small and medium-sized enterprises (SMEs) as a programme of the Federal Ministry for Economic Affairs and Climate Action.

References

1. Friedrich, K.: Carbon fiber reinforced thermoplastic composites for future automotive applications: AIP Conference Proceedings
2. Bader, B., Türck, E., Vietor, T.: Multi material design. A current overview of the used potential in automotive industries. In: Dröder, K., Vietor, T. (eds.) Technologies for economical and functional lightweight design. Berlin, Heidelberg (2019)
3. Bader, B., Berlin, W., Demes, M.: Interdisciplinary research for the development and realization of a structural component in multi-material design suitable for mass scale production. In: Dröder, K., Vietor, T. (eds.) Technologies for economic and functional lightweight design. ZL, pp. 17–24. Springer, Heidelberg (2021). https://doi.org/10.1007/978-3-662-62924-6_2
4. Spiegel, A., Wafzig, F., Giehl, S., Fehrenbacher, U.: Multi-Material-Systeme für Karosserie-Außenhautbauteile von Nutzfahrzeugen. ATZ Automobiltech. Z. **118**(5), 46–51 (2016)
5. Wambua, P., Ivens, J., Verpoest, I.: Natural fibres: can they replace glass in fibre reinforced plastics? Compos. Sci. Technol. **63**(9), 1259–1264 (2003)
6. Babu, T.N., Shyam, S., Kaul, S., Prabha, D.R.: Natural fibre composites – an alternative to plastics in the automotive industry: A review. Proceedings of the Institution of Mechanical Engineers, Part L: J. Mater. Des. Appl. **236**(2), 237–236 (2022)
7. Nunes, B., Bennett, D.: Green operations initiatives in the automotive industry. Benchmark. Int. J. **17**(3), 396–420 (2010)
8. Bodig, J., Jayne, B.A.: Mechanics of wood and wood composites, Repr. Ed. Malabar, Fla. (1993)
9. Wagenführ, A.: Die strukturelle Anisotropie von Holz als Chance für technische Innovationen. Stuttgart (2008)
10. Böhm, S., Kohl, D.: Verbundprojekt: Holzformteile als Multi-Materialsysteme für den Einsatz im Fahrzeug-Rohbau, Akronym: HAMMER: Teilprojekt: Integration von holzbasierten Multimaterialsystemen in Fahrzeugstrukturen durch geeignete Fügetechnologien und in Rohbaufertigungen durch geeignete Prozessketten: Schlussbericht zum Vorhaben: Laufzeit: 01.04.2012 bis 30.04.2016 (2016)

11. Zerbst, J. D.: Entwicklung einer virtuellen Prozesskette zur rechnergestützten Simulation der Umformung von textilkaschierten Holzoberflächen dekorativer Bauteile im Fahrzeug-Innenraum, Dissertation. Dresden (2022)
12. Reichel, V., Berlin, W., Rothe, F., Beuscher, J., Dröder, K.: Study on one-shot process for wood-based composites (2020)
13. Dipl.-Geogr. Dominik Vogt, Dipl.-Phys. Michael Karus, Dipl.-Ök. Sven Ortmann, Dipl.-Ing. Christin Schmidt, Dipl.-Gwl. Christian Gahle: Studie Wood-Plastic-Composites (WPC) Holz-Kunststoff-Verbundwerkstoffe. Märkte in Nordamerika, Japan und Europa mit Schwerpunkt auf Deutschland Technische Eigenschaften – Anwendungsgebiete Preise – Märkte – Akteure (2006)
14. Klyosov, A.A.: Wood-plastic composites (2007)
15. Enzensberger, W.: Moderne Beschichtungsverfahren für Holzwerkstoffplatten. Holz als Roh- und Werkstoff **27**(12), 441–63 (1969)
16. Müller, U., Feist, F., Jost, T.: Holzverbundwerkstoffe im Automobilbau der Zu- kunft? – Crashtests und Simulation von Holzwerkstoffen vorausgesetzt!
17. Bäder, E., Koert, H.: Holz+ Kunststoff= Polymerholz. Moderne Holzverarb. **4**(68), 226 (1968)
18. Boehme, C., Hilbert, T.: Formaldehydfrei verleimte Spanplatten, hergestellt mit PVAC-Klebstoffen (Thermoplast-Klebstoffen). Holz Roh- Werkstoff **46**(6), 232 (1988)
19. Buchelt, B., Siegel, C., Wagenführ, A., Nendel, W.: Veneer prepreg - biobased prepregs for thermoplastic processing methods. J. Plast. Technol. **11**(6), 356–374 (2015)
20. Klemt, C.: Verfahrensentwicklung zur Einbringung endlosfaserverstärkter Thermoplaste in metallische Strukturen mittels Patchen, Dissertation. Dresden (2016)
21. Stadlmann, A., Veigel, S., Dornik, F., Pramreiter, M., Steiner, G., Müller, U.: Bond strength of different wood-plastic hybrid components prepared through Back injection moulding. BioResources **15**(1), 1050–1061 (2020)
22. Kuusipalo, J.: Plastic coating of plywood using extrusion technique. Silva Fennica **35**(1), 103–110 (2001)
23. Reichel, V., Berlin, W., Rothe, F., Beuscher, J., Dröder, K.: Study of shear-cutting mechanisms on wood veneer. Forests **11**(6), 703 (2020)
24. Rowell, R.M., Pettersen, R., Han, J.S., Rowell, J.S., Tshabalala, M.A.: Cell wall chemistry. Handbook of wood chemistry and wood composites **2**, 33–72 (2005)
25. Bertoniere, N.R., Zeronian, S.H.: Chemical characterization of cellulose (1987)
26. Maréchal, Y., Chanzy, H.: The hydrogen bond network in I β cellulose as observed by infrared spectrometry. J Mol. Struct. **523**(1–3), 183–196 (2000)
27. Rezaee Niaraki, P., Krause, A.: Correlation between physical bonding and mechanical properties of wood–plastic composites: Part 2: Effect of thermodynamic factors on interfacial bonding at wood–polymer interface. J. Adhes. Sci. Technol. **34**(7), 756–768 (2020)
28. Osswald, T., Rudolph, N.: Polymer rheology. Fundamentals and applications. Munich (2014)
29. Whiteley, K.S., Heggs, T.G., Koch, H., Mawer, R.L., Immel, W.: Ullmann's encyclopedia of industrial chemistry. Chichester (2010)
30. Maier, R.D., Schiller, M.: Handbuch Kunststoff-Additive, 4., completely new revised edn. München (2016)
31. Takase, S., Shiraishi, N.: Studies on composites from wood and polypropylenes. II. J Appl Polym Sci **37**(3), 645–59 (1989)
32. Bastian, M., Radovanovic, I.: Einfluss der Haftvermittler. Kunststoffe **8**, 49–53 (2005)
33. Erdmann, M.: Entwicklung innovativer Compatibilizer-Systeme für Polypropylen-Nanocomposites auf der Basis organophiler Schichtsilikate. Berlin (2009)

34. Niemz, P., Sonderegger, W.U.: Holzphysik. Physik des Holzes und der Holzwerkstoffe, München (2017)
35. Wagenführ, A., Scholz, F. (eds.): Taschenbuch der Holztechnik. München (2018)
36. Dröder, K. (ed.): Prozesstechnologie zur Herstellung von FVK-Metall-Hybriden. ZL, Springer, Heidelberg (2020). https://doi.org/10.1007/978-3-662-60680-3
37. Barth, S., Hartmann, M., Rudolph, M., Karlinger, P., Michanickl, A., Schemme, M.: Zusammen Massnehmen - Verfahrensenticklung zum Hinterspritzen von Holzfurnieren am Beispiel eines Zillstocks. In: Kunststoffe (2016) 7

Generative Manufacturing

Atmospheric Pressure Plasma Sources for Additive Manufacturing

Thomas Neubert(✉), Kristina Lachmann, Lara Schumann, Veysel Zeren, Tim Abraham, and Michael Thomas

Fraunhofer Institute for Surface Engineering and Thin Films IST, Bienroder Weg, 54E, 38108 Braunschweig, Germany
{thomas.neubert,kristina.lachmann,lara.schumann, veysel.zeren,michael.thomas}@ist.fraunhofer.de, tim.abraham@tubraunschweig.de

Abstract. Adhesion plays a central role in additive manufacturing (AM) processes like Fused Deposition Modelling (FDM). The printed part needs sufficient adhesion to the build plate. The adhesion in between the printed material layers defines the mechanical strength of the component. Post-processing processes such as painting or bonding require good chemical coupling, as does the combination of different materials (FDM of different polymers, integration of metal inserts).

Unfortunately, sufficient adhesion is not given for all material combinations. An interesting and flexible possibility to influence the surface chemistry and thus the chemical bonding are surface functionalization by using atmospheric pressure plasma technologies. These have been used for many years with great success for the surface treatment of polymers. Depending on the plasma parameters used, it is possible to clean and etch surfaces, to functionalise surfaces with reactive chemical groups, to deposit layers or to cross-link polymers. While this is already used for treatment of 2D and 3D substrates in many applications, the integration of new functionalities by plasma into additive manufacturing processes is of great interest. This allows new functionalities to be integrated into 3D-printed components and simplifies further processing.

At the Fraunhofer IST various approaches of small atmospheric pressure plasma sources are under development which can be easily integrated into the additive manufacturing process. Small point-shaped plasma modules for the sequential treatment of defined areas were built-up and integrated into commercially available FDM printers. With these plasma sources, the plasma treatment of defined areas of the components is possible during 3D printing. Integrated plasma sources enable moreover the in-situ treatment of internal surfaces which are subsequently no longer accessible for other treatment processes. This can create new products and expand the scope of additive manufacturing.

Keywords: Atmospheric plasma source · Additive Manufacturing · Adhesion · Polypropylene

K. Dröder and T. Vietor (Eds.): *Future Automotive Production Conference 2022*, Zukunftstechnologien für den multifunktionalen Leichtbau, pp. 147–155, 2023.
https://doi.org/10.1007/978-3-658-39928-3_11

1 Introduction

The surface chemistry influences adhesion to a considerable extent. A common way to optimize the surface chemistry is the functionalization by using atmospheric pressure plasmas for increasing the adhesive strength of paintings and adhesives on polymer surfaces [1, 2] or between different polymers [3]. The plasma serves as a source of energy to trigger various reactions to generate different physical and chemical processes on the surfaces. Organic impurities can be oxidized and removed, organic surfaces can be etched and roughened, chemical functionalisation can be achieved or even layers with desired functions can be deposited [4].

A frequently used possibility to generate atmospheric pressure plasmas are so-called dielectric barrier discharges (DBD). Here, a high alternating voltage is applied to two electrodes. Between the electrodes there is at least one dielectric barrier and a gas gap. The high electric fields cause an electric gas discharge, the currents of which are limited by the dielectric barrier. Such plasma sources are available in a wide variety of designs and sizes, including plasma nozzles (plasma jets) that can be used robot-guided for the treatment of surfaces [5–7]. Such a plasma nozzle can be integrated into an FDM 3D printer by mounting them parallel to the print head [8]. This makes it possible to plasma treat the printed structure before, during and after the printing process. It is also possible to treat different surfaces of the printed component differently to create gradients. Many commercial plasma nozzles are relatively large and structurally difficult to integrate into an FDM 3D printer. Our approach is to develop particularly small, robust and flexible plasma sources for integration in AM processes. This allows a cost-effective implementation to improve the mechanical stability of printed polymer components, to ensure further processability (bonding, painting) and to enable adhesive combinations of different materials (polymer composites, metal inserts).

2 Experimental

In our experiments a small plasma nozzle based on a μ-DBD was developed (Fig. 1). The plasma nozzle has external dimensions of $37 \times 20 \times 10\ \text{mm}^3$, is made of alumina, brass and copper and weighs approx. 15 g. In principle, the microplasma nozzle consists of a 3D-printed ceramic tube (inner diameter 7 mm) made of alumina with a central high-voltage electrode and a grounded outer tube as counter-electrode. Similar designs are mentioned in [6] and [7]. The outlet for the ionised gas is narrowed to achieve a high gas velocity at the tip of the nozzle. Helium (3 slm) was used as process gas to ignite the plasma and create a plasma plume of >10 mm. A Plasmagenerator G2000 from Redline Technologies was used to power the plasma source. The expected advantage of the chosen design is the small size and mass, the simple construction and the high stability.

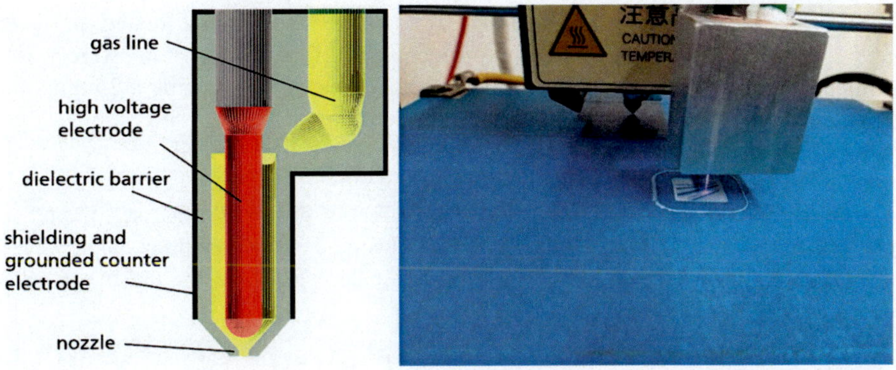

Fig. 1. Left: Construction of the used plasma nozzle / right: Experimental setup with plasma nozzle mounted next to extruder nozzle in an FDM printer

The plasma source is mounted on a commercial FDM 3D printer of the type Anicubic i3 Mega S. To control the movement of the plasma source and the printing process G-code files were generated with the software Cura from Ultimaker. Separate G-code files were generated for the printed polymer structure and the area which is to be plasma treated. Both files are then combined into one G-code file by script in order to control both the polymer deposition and the plasma treatment layer by layer.

Two different polymers were tested. Commercial 1.75 mm filaments made of polylactic acid (PLA) from Polylite™ and polypropylene (PP) from BASF/Innofil3D were used.

An important tool for evaluating possible adhesion is the determination of the surface energy. To analyse the treated and untreated surfaces contact angle measurements for water and diiodomethane were performed by using the "Mobile Surface Analyser" from Krüss GmbH to measure the surface energy of plasma treated polyethylene (PE) foil. By using the method of Owens, Wendt, Rabel and Kaelble (OWRK) the surface energy as well as its polar and dispersive component can then be determined [9]. However, for the 3D-printed surfaces, this method delivers inaccurate results because of the roughness of the surfaces. Instead, test inks from the company Tantec A/S were used to evaluate tendencies for the surface energy. This set of test inks with varying surface tensions wets surfaces with different surface energies differently [10].

To test the adhesive strength of lacquers on plasma-treated printed polymers, $25 \times 25 \times 1\ mm^3$ cuboids were first printed. Subsequently, the entire upper surface of the samples was plasma-treated and lacquered. A commercial water-based spray paint of the type "Dupli colour aqua lacquer" from Motip Dupli GmbH was used for this. After curing, cross-cut tests according to DIN EN ISO 2409 were carried out with adhesive tape to evaluate the adhesion of the lacquer.

In order to investigate the bonding strength on the plasma-treated polymers, 2 cuboid test specimens each with the dimensions 25 × 10 × 1 mm^3 were printed and bonded after plasma treatment with a contact area of 1 cm^2 using a 2-component epoxy adhesive type Loctite® EA3430 from Henkel AG. After curing, the test specimens of a tensile testing machine type 81816 from Karl Frank GmbH were pulled apart and the maximum shear force was determined.

In order to be able to compare the plasma effect of different processes, the plasma dose was determined according to Eq. 1. The plasma dose D is defined as the amount of energy E introduced per area A. This is then calculated from the plasma power P, the scan width s and the scan speed v.

$$D = \frac{E}{A} = \frac{P}{s \cdot v} \tag{1}$$

3 Results

In order to determine the effect of the treatment of the μ-plasma nozzle on the wettability and the surface energy, polyethylene films were treated. For this purpose, the μ-plasma jet was flown through with 3 slm helium and operated at 6 W power, 72 kHz and 3.4 kV. The plasma jet was moved at a distance of 3 mm to the surface with a line width of 0.4 mm at speeds of 50 mm/s, 25 mm/s, 12.5 mm/s and 6.25 mm/s. The treated PE films were analysed by contact angle measurements using the OWRK-method. The determined water contact angles and the surface energy with its disperse and polar components are shown in Figs. 2 and 3, respectively. It was observed that there is already a significant lift in the wettability of the PE surface at relatively low plasma doses of 0.4 J/mm^2 (corresponding to 6 W power at 50 mm/s scanning speed). The change in the total surface free energy (SFE) from 32 to 54 mN/m is mainly due to the change in the polar component from 0 to 15 mN/m, indicating the formation of polar groups on the PE surface. Thus the used plasma source already causes a significant change in the surface chemistry of the PE at typical printing speeds of FDM 3D printers.

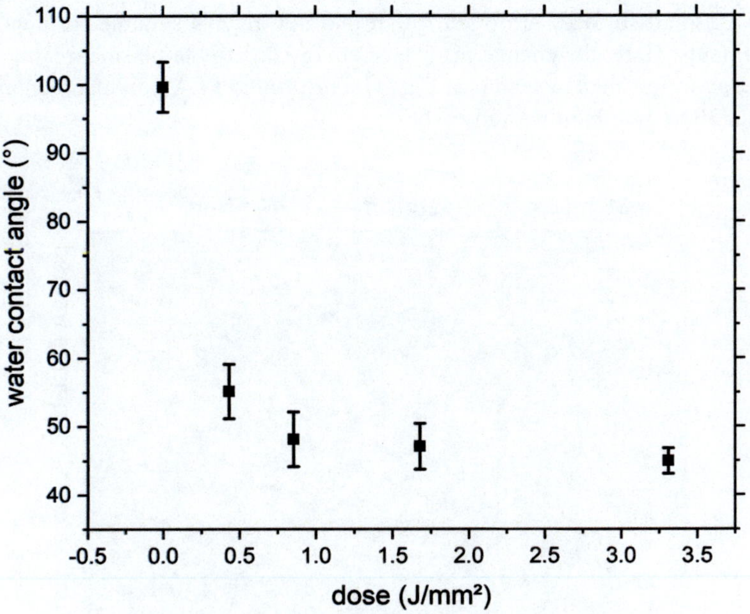

Fig. 2. Water contact angle on PE for different plasma doses

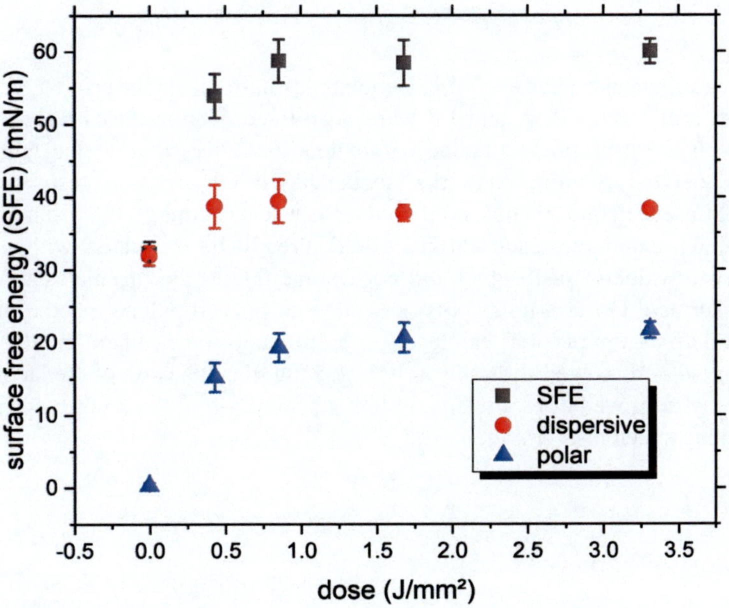

Fig. 3. Surface free energy (SFE) with its polar and dispersive fractions as a function of plasma dose of treated PE films

FDM-printed surfaces are often rough and not always suitable for contact angle measurements. Here, the change in surface energy due to the plasma treatment can be evaluated with test inks instead (see Fig. 4). For printed PLA polymer a change of the SFE from 30 to 44 mN/m was observed.

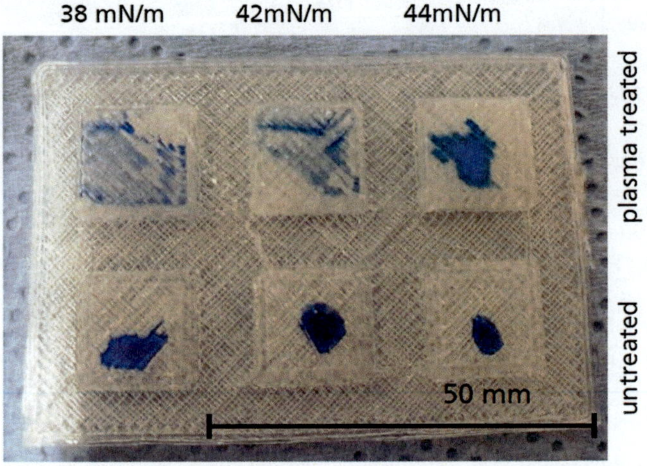

Fig. 4. 3D-printed PLA surface with plasma-treated (top) and untreated (bottom) areas wetted with test inks

To investigate the adhesion of a lacquer plasma-treated 3D printed, cuboids (25 × 25 × 1 mm^3) from PLA and PP were 3d-printed. The surfaces were subsequent treated with the μ-plasma jet (same parameters as above) and coated with a water-based lacquer by spraying. After the lacquer has dried, cross-cut tests were carried out. No difference could be observed on the PLA surface due to the plasma treatment. Here, the untreated reference surface could already be evaluated with a cross-cut characteristic value of 0–1, which did not change due to the plasma treatment of the polymer surface. On non-polar polymers such as polypropylene, on the other hand, atmospheric pressure plasmas can lead to a significant improvement in adhesion. This was observed here as well. It is shown in Fig. 5 that the adhesion of the lacquer can be significantly improved after a relatively low plasma dose of 0.4 J/mm^2, corresponding to a scanning speed of 50 mm/s.

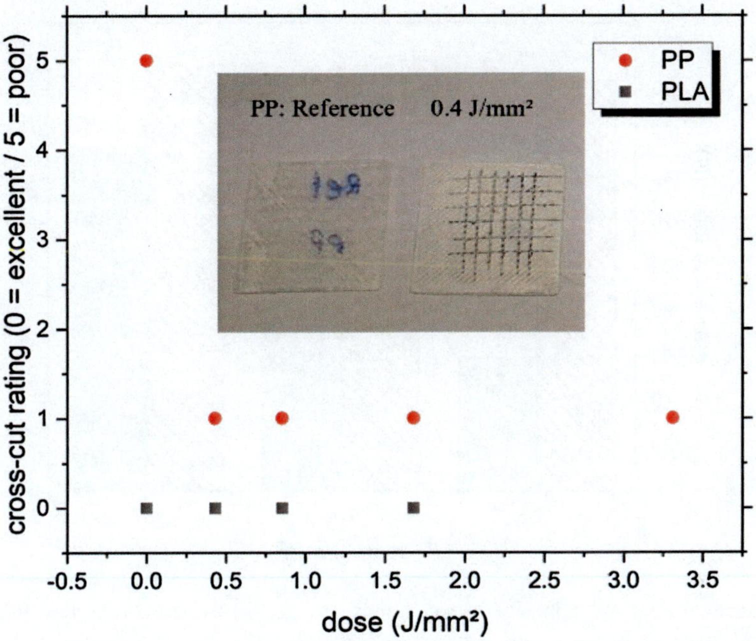

Fig. 5. Results of the cross-cut ratings on plasma-treated and lacquered 3D-printed PLA and PP surfaces

As second application bonding of two 3D printed polymer surfaces was investigated. To determine the strength of the adhesive bond, cuboids of PLA and PP were plasma-treated 3D printed and subsequently the surface was plasma treated with different doses and bonded with a 2-component epoxy resin. After curing, the parts were pulled apart parallel to the 1 cm^2 bonding surfaces and the shear forces were determined (see Fig. 6). A very similar behaviour to the adhesion of the lacquer was observed. On the PLA surfaces, a high adhesion already occurred without plasma treatment, which did not change further with the plasma treatment. On the untreated PP, on the other hand, only a low adhesive strength of the adhesive was observed, which could increase considerably to the level of the PLA by a factor of 2.5 through plasma activation (i.e. through the generation of polar groups).

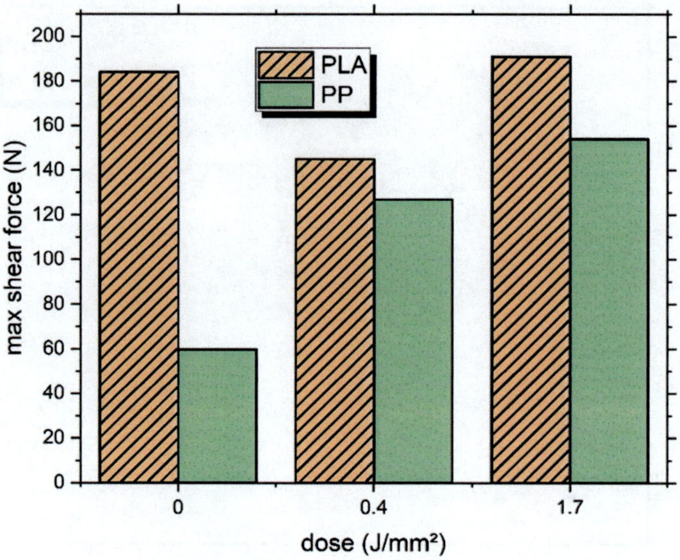

Fig. 6. Shear forces of adhesive epoxy-bonds of plasma-treated PLA and PP samples ($A = 1\ cm^2$)

4 Conclusions

In this work we were able to demonstrate that small stable plasma sources can be integrated into commercially available FDM 3D printers. A particularly small ($37 \times 20 \times 10\ mm^3$), robust (alumina, brass, copper) and light (15 g) plasma nozzle was developed, which could be easily integrated into a commercial FDM 3D printer. With this plasma nozzle it was possible to create functional groups on the surface of 3D printed polymers and thus increase the surface energy. The adhesion on the printed polymers can be greatly improved, especially on non-polar polymer surfaces. This could be shown in this work for polypropylene. It was possible to improve the adhesive strength of lacquer coating from a cross-cut characteristic value of 5 to 1 and to improve the shear strength of adhesive bonds by a factor of 2.5. The treatment speed required for this is 12 to 50 mm/s, which is within the typical range for FDM 3D printing processes.

Since such treatments are also possible on the inner surfaces of the printed components, which are not accessible to subsequent plasma activation, this opens up a number of new interesting fields of application, especially for the combination of different types of materials. The expected advantage of the chosen design for the plasma nozzle (small size and mass, the simple construction and the high stability) could be confirmed. It is particularly advantageous that the alumina base body of the plasma nozzle itself was manufactured by means of additive manufacturing and thus offers great design freedom for further developments. Reasonable further developments of the investigated technology concern the use of other process gases (especially air). This makes the set-up independent of special process gases. Furthermore, we are working on ring-shaped plasma sources that allow parallel treatment of the printed surfaces and thus save considerable processing time and costs. Further modifications of the source should allow plasma-enhanced chemical vapour deposition (PECVD) processes. With this, migration barrier layers, coatings to promote or inhibit cell growth or electrically conductive coatings in the printed polymer components can be achieved in addition to further modifications of the surface energy.

References

1. Kehrer, M.: Cold atmospheric pressure plasma treatment for adhesion improvement on polypropylene surfaces. Surf. Coat. Technol. **403**, 126389 (2020)
2. Frascio, M.: Appraisal of surface preparation in adhesive bonding of additive manufactured substrates. Int. J. Adhes. Adhes. **106**, 102802 (2021)
3. Boros, R.: Plasma treatment to improve the adhesion between ABS and PA6 in hybrid structures produced by injection over molding. Polym. Testing **106**, 107446 (2022)
4. Tendero, C.: Atmospheric pressure plasmas: A review. Spectrochim. Acta Part B **61**, 2–30 (2006)
5. Arnold T.: Precision asphere and freeform optics manufacturing using plasma jet machining technology. In Proceedings of SPIE Volume 10448 on Optifab 2017, 1044814 (2017)
6. Fanelli, F.: Atmospheric pressure non-equilibrium plasma jet technology: general features, specificities and applications in surface processing of materials. Surf. Coat. Technol. **322**, 174–201 (2017)
7. Winter, J.: Atmospheric pressure plasma jets: an overview of devices and new directions. Plasma Sources Sci. Technol. **24**, 064001 (2015)
8. Narahara, H.: Improvement and evaluation of the interlaminar bonding strength of FDM parts by atmospheric-pressure plasma. Procedia CIRP **42**, 754–759 (2016)
9. Owens, D.K.: Estimation of the surface free energy of polymers. J. Appl. Polym. Sci. **13**, 1741–1747 (1969)
10. Aydemir C.: Surface analysis of polymer films for wettability and ink adhesion. Color Res. Appl. 1–11 (2020)

Design Freedoms of Lattice Structures for Interlock Bonding

Raphael Freund[1](✉), Fynn Matthis Sallach[1], and Thomas Vietor[2]

[1] TU Braunschweig, Institute for Engineering Design, Langer Kamp 8, 38106 Braunschweig, Germany
{r.freund,f.sallach}@tu-braunschweig.de
[2] TU Braunschweig, Institute for Engineering Design, Hermann-Blenk-Straße 42, 38108 Braunschweig, Germany
t.vietor@tu-braunschweig.de

Abstract. Additive manufacturing (AM) enables new design freedoms that are often not fully utilized for complexity and price reasons. Lattice structures are one use case of AM that, despite being frequently utilized in lightweight design, is not commonly used in other areas. Bonding low adhesion materials through interlocking lattice structures uses the design freedom of AM in a new and innovative way, however the design of these structures is highly complex and the mechanical behavior is not easily understood. In order to completely tap into design potentials of such structures, it is first necessary to understand the design freedoms of lattice structures in general to then apply this knowledge to bonding by interlocking lattices specifically. This publication aims to show how lattice structures are designed in a fast and repeatable manner in a CAD environment and tries to show the design freedoms of these lattice structures. These insights are then applied to the use case of bonding by interlocking. It is shown that by using interlocking lattices, the bonding strength of bonds for low adhesion polymers can be significantly increased while achieving low standard deviations, even compared to established methods of increasing bonding strength like plasma pre-treatment.

Keywords: Additive manufacturing · Design for additive manufacturing · Lattice structures · Bonding

1 Introduction

Contrary to subtractive manufacturing, AM processes fabricate a part by adding material layer-by-layer. Thus, they offer new design freedom especially considering complex geometric shapes. In addition, multi-material AM, for example by use of material extrusion with multiple extruders, opens new pathways to integrate material-specific functions, like electrical conductivity or elasticity, without the need for additional joining.

K. Dröder and T. Vietor (Eds.): *Future Automotive Production Conference 2022*, Zukunftstechnologien für den multifunktionalen Leichtbau, pp. 156–169, 2023.
https://doi.org/10.1007/978-3-658-39928-3_12

However, bad adhesive properties between different materials often weaken the compound structure or even make manufacturing impossible in the first place. Nonetheless, the freedom in geometry and material composition offered by AM can be used to facilitate the use of multiple material parts in most cases. If the desired combination of materials show bad adhesion, interlocking structures such as lattice structures can be used to change the failure mode from adhesive failure in the interface to cohesive failure inside the structure, thereby significantly increasing the maximum load.

In addition, these same structures can be used to facilitate bonding by adhesive in cases where adhesion to one adherend is not sufficient. By creating a lattice that can be fully engulfed by the adhesive, the problem becomes one of cohesion in the adhesive rather than adhesion on the interface. This can be used to make materials like polypropylene (PP) that were formerly hard to bond easy to handle. This way, bonding of AM parts becomes possible to circumvent the size restrictions of the manufacturing machines or combine AM with conventional manufacturing processes, in order to save costs. Since the mechanism of bonding of such structures is mainly given by their geometry, their application is largely independent of the manufacturing process or the used material. Even though this publication focuses of material extrusion, similar structures could be manufactured using other AM processes like selective laser sintering or selective laser sintering. Additionally, since the adhesion properties of one adherent become irrelevant, the adhesive can be chosen according to the non-AM part. It can even be conceived that the process be adjusted for metal parts by replacing the adhesive by by solder.

However, the behavior of such lattice structures is highly dependent on their design. In order to fully utilize the potentials of lattice structures for technical applications, their design freedoms need to be understood thoroughly. Hence, this publication examines the different freedoms in designing such interlocking structures, and shows their application as a way to combine materials with bad adhesive properties into a load bearing structure.

2 Mechanical Interlocking Structures by Material Extrusion

Material extrusion (MEX, sometimes also FFF or FDM) is an AM process that uses heating and plastification of thermoplastic polymer pre-product to fabricate a part layer by layer. Manufacturing costs of MEX parts are independent of the geometric complexity, but mechanical properties are lower and highly anisotropic. [1] Switching between different extruders during fabrication makes combination of different materials in the same part and even the same layer possible, which gives way to manufacturing of multi-material parts without additional processing steps. [2] A schematic depiction of MEX is shown in Fig. 1.

Different categories of multi material AM consist of discrete multiple materials, composite materials and porous materials. [3] This publication focuses on discrete material transitions as a way to increase performance of multi material parts or adhesion bonds. Use of materials with different properties like hardness, temperature resistance or electrical conductivity is an effective means for integration of functions. [4–8]

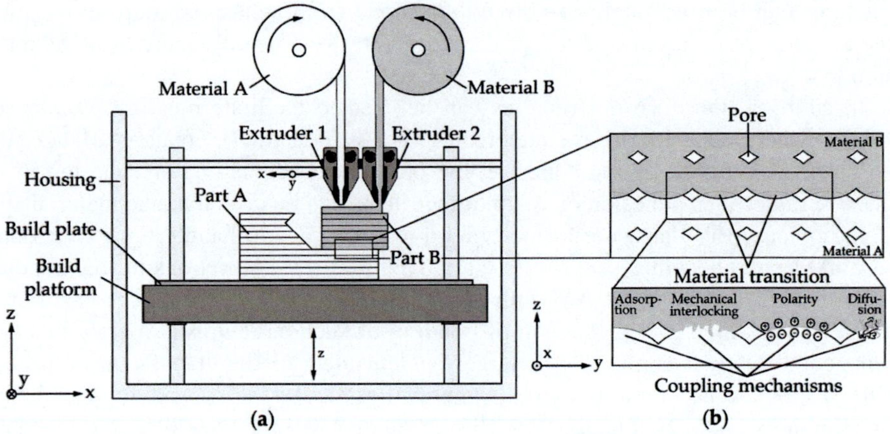

Fig. 1. (**a**) Schematic representation of material extrusion principle and (**b**) material transition including relevant coupling mechanisms. [9]

Mechanical properties of multi material parts manufactured in this manner highly depend on the adhesive properties of the interface. These properties themselves are influenced not only by the materials used, but also by the process parameters during printing. Previous examinations show that it is possible to influence composite strength by choice of material couples and process parameters but also by incorporation of mechanically interlocking structures. [10]

Lattice structures consist of a regular repeating pattern of unit cells consisting of plates or struts and are uniquely suited to form such mechanical interlocks. Usually they have high-performance over weight ratio [11], high energy absorption [12] and unique acoustic properties [13]. In addition, they offer a high degree of design freedom, which allows for graded structures with fine-tuned mechanical behavior. Because of the high geometric complexity of these structures, they are exclusively manufactured with AM technologies. In addition, these structures can be combined with multi material AM in the form of mechanical locks for increasing bonding strength and can even outmatch conventional ways of increasing adhesion like plasma treatment. [14]

However, the behavior of these structures is not well understood, since they are highly anisotropic and can range in sizes close to the resolution limit of most manufacturing machines.

3 Design of Lattice Structures Using Grasshopper Rhino

Several existing tools can generate lattice structures in a CAD environment. Autodesk within, nTopology, PTC Creo and many more show the functionality to easily incorporate lattice structures into CAD parts. However, those tools are sometimes restricted to certain kinds of lattice structures and the outcomes are not always 100% predictable. Hence, this publication uses Grasshopper 3D [15], a visual programming language for the CAD application Rhinoceros 3D, in order to create stl-files for printing. Grasshopper is often used for parametric modelling in structural engineering or architecture, but is used here for its open nature that allows adjustment of all possible parameters in a lattice.

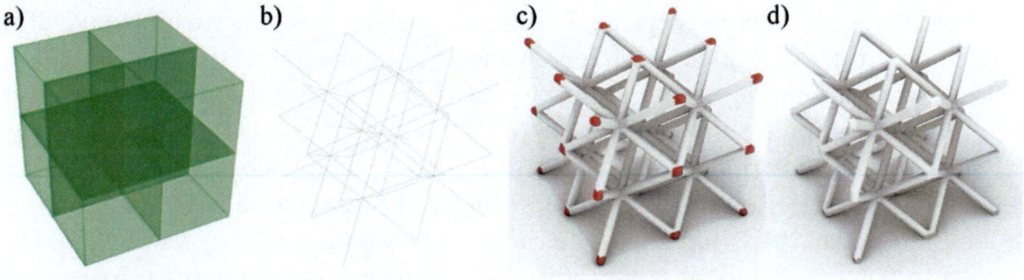

Fig. 2. Lattice generation with the Grasshopper script. **a)** voxelization, **b)** population, **c)** thickening, **d)** trimming

The developed design chain can be seen in Fig. 2. It is fully parametric and only needs user input regarding the main design values of the lattice. The script can generally be divided into 4 steps, voxelation, population, thickening and trimming. It first creates a bounding volume either in Grasshopper itself or by importing an stl-file into the script. After defining the base plane and thereby the orientation of the lattice, the volume is subdivided into voxels based on the desired cell size gradients. These boxes are filled with unit cells in the next step. For this, a unit cell consisting of line segments has to be defined first either by choosing a pre-made structure or manually creating a structure from scratch. Commonly used unit cells are already present in plugins for Grasshopper such as Crystallon, however for creating more complex or less symmetric cells they have to be generated manually. Next, the unit cell is morphed to neatly fit into each seperate box created earlier. This results in a collection of centerlines of all struts in the lattice that have to be thickened in the last step by subdivision into points that serve as center-points of spheres. The radius of those spheres is variable and can be input according as graphical or formulaic expression or via importing values, for example based on a topology optimization. Creation and addition of these spheres is achieved using the plugin Dendro, which is developed by ECR.Labs and built upon the volume data structure in the OpenVDB library. This facilitates quick computation of highly complex lattice structures with high repeatability. [16] Since the centers of these spheres are very close to each other, the result is an almost completely smooth structure with surface deviation lower than the resolution of the used AM machines.

Since the thickening process of the points on the boundary creates structures, that slightly protrude out of the design volume, the structure has to be cut using a Boolean operation. In the last step this structure is converted into an stl-file and exported for printing.

4 Design Freedom of Lattice Structures

In order for product developers to design interlocking lattice structures and utilize them in a useful way the design freedoms need to be demonstrated first. Since most lattice structures consist of an arrangement of struts it is useful to first evaluate them according to their unit cell.

The Maxwell criterion predicts the stiffness of structures consisting of nodes and trusses. For example, a system is under-stiff, i.e. bending-dominated, if the ratio of struts (s) to nodes (n) is too small, i.e. its Maxwell number (M) given by Eq. (1) is lower than 0. Accordingly, it is just-stiff if M is exactly 0 or over-stiff if M is greater than 0 and shows stretch-dominated behavior

$$M = s - 3n + 6 \tag{1}$$

Under-stiff structures show high compliance and low strength, whereas over-stiff structures show very low compliance and high strength due to a reinforcement effect of the redundant struts. [17] Most technical uses of lattice structures aim for an over-stiff structure. Using the Maxwell criterion, the majority of commonly used cubic unit cells can be shown to be under-stiff. However, once the criterion is applied to a periodic lattice and not only a single cell, many structures can be determined to be over-stiff, since they show at least three times the amount of struts than vertices averaged over the whole lattice. However, in addition to the Maxwell criterion, lattices should also show a connectivity of at least 12 in order to be as stiff as possible. [18]

Hence, in order to achieve stretch-dominated behavior and low density, it is beneficial to construct the unit cell in a way, which maximizes the number of struts by simultaneously minimizing the number of vertices. All following considerations will only regard cubic lattices, since they are easiest to create and visualize. Nonetheless, the results can be transferred to lattices with different symmetries.

There are in total 32 possible struts (12 edges, 12 face diagonals and 8 body diagonals) and 15 possible vertices (8 corners, 6 face centers, 1 body center) in a cubic lattice as shown in Fig. 3. Adding these numbers for an infinite lattice results in a ratio of 32 struts for every 5 vertices.

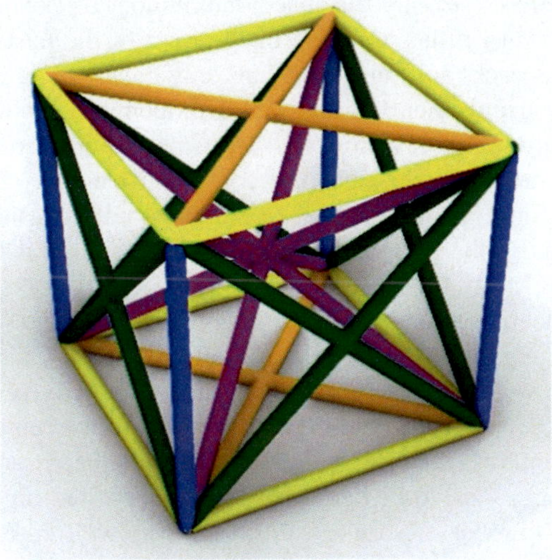

Fig. 3. Different types of struts in cubic lattices: horizontal edges (yellow), vertical edges (blue), horizontal face-diagonals (orange), vertical face-diagonals (green), body diagonals (purple)

Other possible connections of vertices, like struts between the body center and a face center or connections of two face centers, will not be discussed in this publication. In almost all cases these lines can be thought of as the edges, face diagonals or body diagonals of a smaller unit cell from which all other structures can be constructed using symmetry operations, as shown in Fig. 4. Since these smaller cells all show the same connectivity because of symmetry, they are identical for this purpose.

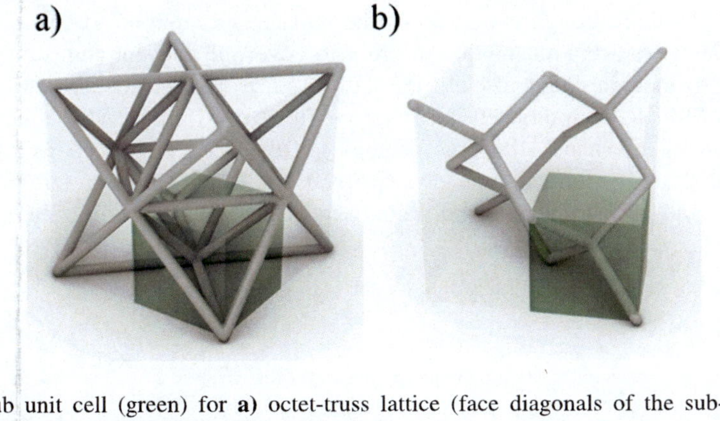

Fig. 4. Sub unit cell (green) for **a)** octet-truss lattice (face diagonals of the sub-lattice), **b)** diamond lattice (body diagonals of the sub-lattice)

Since the lattice has to evenly tile space, the number of existing corner vertices in the unit cell or sub unit cell has to be an integer part of the maximum possible number of 8. This means each (sub) unit cells can only have 8, 4 or 2 corner vertices (all corners, tetrahedral arrangement or diagonal arrangement). Following Fig. 3 for the case of 8 corners vertices, a maximum of 4 struts per corner vertex are possible when averaging over all corners. In the case of 4 corner vertices only 2.25 struts per vertex on average remain. Since a lattice with only 2 corners only consists of body diagonals in one direction and can therefore not be three-dimensional, it is not further discussed in this publication.

Regarding connectivity, a lattice with 8 corners can have a maximum connectivity of 20 for the corners, 8 for the body diagonals and 4 for the face diagonals. Accordingly, lattices with 4 corners can show a connectivity of no more than 18 at the corners or 4 at the body diagonals. This shows that face center vertices should be avoided for stiff structures, because they do not show the necessary connectivity.

Body center vertices are to be avoided as well albeit less urgently, since they can at least add a large number of struts to fulfill the Maxwell criterion. When omitting all struts that would result in face center vertices, an 8-cornered cubic cell can have a maximum of 26 struts and 7 vertices, since one strut on each face is omitted. Counting these vertices and faces for an infinite lattice results in a ratio of 1:7, enough to fulfill the Maxwell criterion of a ratio of 1:3. Additionally cutting all but one body diagonal and therefore only using corner vertices, results in 19 struts and 8 vertices with a ratio of 1:7 as well. This shows that the negative influence of body center vertices is way less severe than that of face center vertices.

Apart from the number and types of struts, orientation and placement can play an important role in the design of lattice structures as well. It is beneficial to orient the lattice in such a way, that most struts are along the load direction. Figure 3 shows the possible strut types in a cubic lattice. While face diagonals and body diagonals show angles of 45° and 35.3° to the horizontal respectively, values close to these angles can be achieved by adjusting the voxel size in one direction. This becomes possible since the voxels of the lattice do not necessarily have to be exactly cubic, making changing the angle possible through distortion of the unit cell in addition to changing its orientation. However, this means that in order to cover all relevant angles, it is usually necessary to include diagonals into the lattice, since edge struts can only ever have an angle of 90° to each other. It is further possible to distort the voxels largely independently into any shape, adjusting the angles of the struts as needed, as long as they still fill the design volume completely. This is useful especially for rotational parts, as the cells could also be constructed radially in this manner, adjusting the strut directions to better carry radial loads.

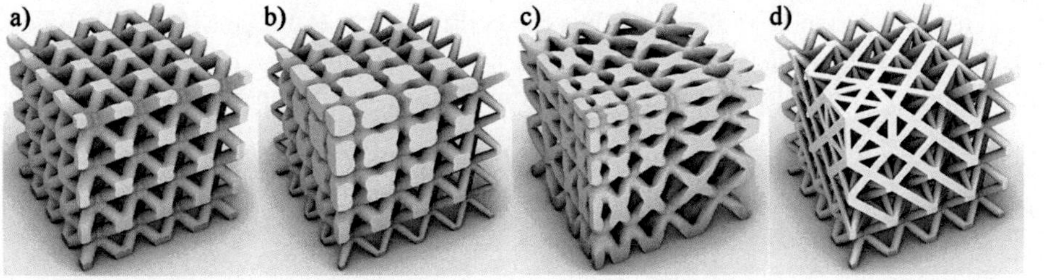

Fig. 5. Density gradients in a 4x4x4 body-centered cubic lattice through different means. **a)** No gradient, **b)** quadratic diameter gradient in all directions, **c)** linear cell size gradient in all directions, **d)** strut type gradient

Additionally, one big advantage of lattice structures is their adjustability. Since stiffness and toughness of these structures depend largely on their density, they can be adjusted in different ways. On the one hand, the density can be changed by changing the diameter of the struts (Fig. 5b) though overall manufacturing constraints have to be taken into account. The minimum strut diameter should not fall below 4 times the nozzle diameter or line thickness of the manufacturing machine, meaning at least two full shell lines are necessary, since smaller diameters lead to excessive warping and increased defect rate. On the other hand, diameters larger than the cell dimensions lead to consolidation of the vertices and struts, which in turn makes the structure almost completely solid, eliminating them from use in certain cases like interlock bonding.

A rule of thumb for adjusting the density in this manner is that for cells with relatively low density, doubling the diameter roughly quadruples the density. This is true, because for small diameters and few vertices the majority of the lattice consists almost exclusively of struts. Since they can be viewed as cylinders with unchanging length, their volume depends on the square of the diameter. Only at the vertices where multiple struts meet, must the structure be approximated by a polyhedron or a sphere, whose volume depends on the diameter cubed.

Adjusting strut diameter is an easy way to include density values from a topology optimization, since the voxel sizes in the lattice generation script can be exactly the same as in the optimization. By not restricting the density values of the optimization to 0 or 1, new possibilities arise for the mechanical properties of the design.

Secondly, the density can be adjusted by varying the voxel size (Fig. 5c). Smaller cells increase the density, while larger cells leave more space and thereby decrease the density. It is important to keep in mind, that changing the cell size also distorts the lattice, which changes the angles of the different strut types. Additionally, the relationship between cell size and density is more complicated.

Since the diameter of the struts stays the same, their volume scales linearly with their length. This means that the volume of cube edges scales linearly with the respective cell dimension, while the volume of face diagonals and body diagonals are proportional to $\sqrt{a^2 + b^2}$ and $\sqrt{a^2 + b^2 + c^2}$ respectively, with a, b and c being the side lengths of the voxel dimension. This means halving the voxel dimensions evenly in x, y and z, roughly leads to quadrupling the density for large voxels with few vertices.

The last influence on the density is the cell type of the unit cell (Fig. 5d). By adding or removing struts from the unit cell, the density can be adjusted. Doubling the number of struts in the unit cell roughly doubles the density in large voxels with few vertices. Additionally, adding struts in areas of major forces in the component preferably in directions close to the direction of said force can greatly increase mechanical properties.

5 Designing Interlocking Lattice Structures

One underutilized case for lattices are interlocking structures as a way increase bonding strength of low adhesion bonds. By utilizing mechanical interlocking low adhesion bonds can be strengthened through cohesive forces in cases where adhesive strength is not sufficient. The goal is for the adhesive to engulf a lattice made of low adhesion material, in order to enable bonding through mechanical locking rather than adhesion. Figure 6 shows a lattice structure used for this purpose in addition to the method of characterizing the tensile strength of the bond. The specimen geometry has an injection channel and a degas channel, in order to facilitate injection of the adhesive into the joining zone and guarantee a high quality of the bond.

Regarding the design of such lattice structures, some things have to be taken into account in accordance to the design freedoms mentioned above. First of all, the dimensions of the lattice in z-direction must be taken into account. Since adhesive bonds are desired to be as space saving as possible, the height of the lattice was chosen to be not greater than 2.4 mm. Since the struts should at least consist of 2 shell lines to avoid warping and the used manufacturing machine has a nozzle diameter of 0.4 mm, the strut diameter has to be at least 1.6 mm. This means, there is not enough space for more than one row of struts in z-direction, since the lattice must still show undercut area for the adhesive to flow into.

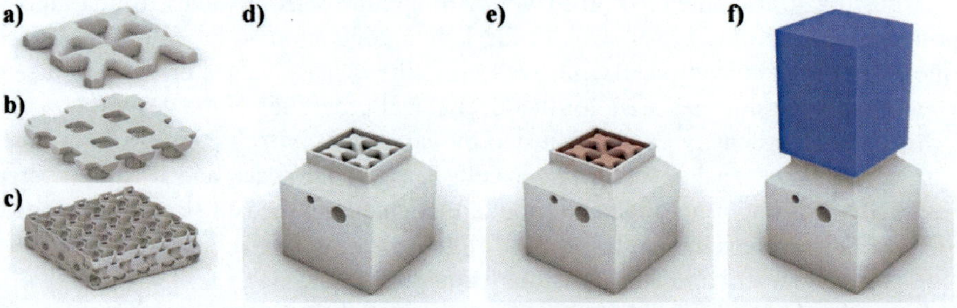

Fig. 6. Specimen geometry for tensile tests. **a)** body centered (BC) lattice geometry, **b)** XYZ lattice geometry consisting of vertical and horizontal struts (XYZ), **c)** gyroid geometry, **d)** printed specimen with BC lattice **e)** lattice engulved by adhesive (red), **f)** Poly(methyl methacrylate) (PMMA) block (blue) for testing.

In order to avoid overlap of the struts and assure that they leave space for the adhesive, their diameter has to be smaller than the lattice voxel size. Reasonable horizontal voxel sizes were estimated to be between 4 and 8 mm, keeping the manufacturing restrictions in mind. In order to leave enough room in between struts for the adhesive to flow into, the strut diameter needs to be significantly smaller than the voxel size however, since not only the toughness of the lattice needs to be considered, but also the toughness of the interlocking adhesive, which occupies the inverse space. In the end, a diameter of 2.4 mm (4 shell lines) proved to be the ideal compromise between strengthening the lattice and leaving enough space for the adhesive to infiltrate the structure.

When choosing the unit cell, the ratio of struts to vertices as well as the vertex connectivity has to be taken into account. Since there is only enough space for one row of struts, the lattice can not be interconnected to a high degree. This means connectivity can be omitted as a criterion and the main goal should therefore be to maximize the ratio of struts to vertices. The most stiff structure is therefore predicted to be a lattice consisting of body diagonals that meet at the center point and horizontal lines further connecting these points. This structure consists of 6 struts and only 2 vertices per cell, when taking the symmetry into account. Vertical struts connecting these points are predicted to have adverse effect, because it would significantly reduce the undercut area for the adhesive to hook into. However, since the interface for the actual adhesive bond needs to be flat, the horizontal lines would need to be only semicircular in cross section reducing their capacity for load bearing. Hence, while horizontal interconnection is predicted to be beneficial for the strength of the lattice, in this case they are expected to even have adverse effects, since they reduce the volume that can be occupied by the adhesive without significant increase in mechanical stength.

Therefore, the first characterized geometry in this publication is the body centered cubic geometry (BC) that is depicted in Fig. 6a). This geometry consist of 4 Unit cells and occupies about 66% of the joining volume. In order to verify the considerations made before, the BC structure is compared to a lattice consisting only of vertical and horizontal struts (XYZ) as seen in Fig. 6b). In order to compensate for the lower mechanical strength of the horizontal struts with only semicircular cross section, the structure was chosen to have 9 unit cells and a slightly higher filling degree of 78%. Lastly, in order to evaluate the potential of surface based lattice as opposed to strut based ones, a gyroid structure with a filling degree of 35% consisting of walls thickness of only a single line width was manufactured, as shown in Fig. 6c). All three structures are fully open-pored, which allows for complete engulfment of the lattice by the adhesive, and avoids non-cohesive failure of the structure.

6 Experimental Setup and Results

All structures were manufactured on an Ultimaker 3 MEX-printer out of PP [19] by Ultimaker, Utrecht, Netherlands, using MEX and bonded onto PMMA blocks using LOCTITE HY 4070 [20] by Henkel AG & Co. KGaAa, Düsseldorf, Germany.

In addition to bonding by interlocking adhesive, one set of specimen was fabricated wherein the adhesive in the joining zone was replaced by polyethylene terephthalate glycol-modified (PETG) using multi-material AM creating a flat surface that was in turn bonded to the PMMA using the same adhesive.

Furthermore, flat specimen without a lattice and without any additional measures as well as flat but plasma pre-treated samples were produced to compare the achievable tensile strengths.

All manufactured structures have been evaluated visually before bonding and show only minor defects mainly in the overhanging areas and due to stringing. While minor warping occurred at the very bottom of the specimens due to low adhesion to the printing bed, it could not be observed in the lattice zone.

As seen in Fig. 7 the tensile strengths of the sample sets with interlocking lattice structures vastly exceed those of conventional adhesive bonding both with and without plasma pre-treatment while also showing low standard deviation. Failure even occurred as mixed fracture in the adhesive and on the interface of the PMMA for all structures, meaning the stiffness of the lattice was sufficient to bear the load in all cases. However, the strength is still significantly lower than the strength of the adhesive on comparable polymers as per data sheet [20], presumably due to weakening of the adhesive layer at the very top of the lattice.

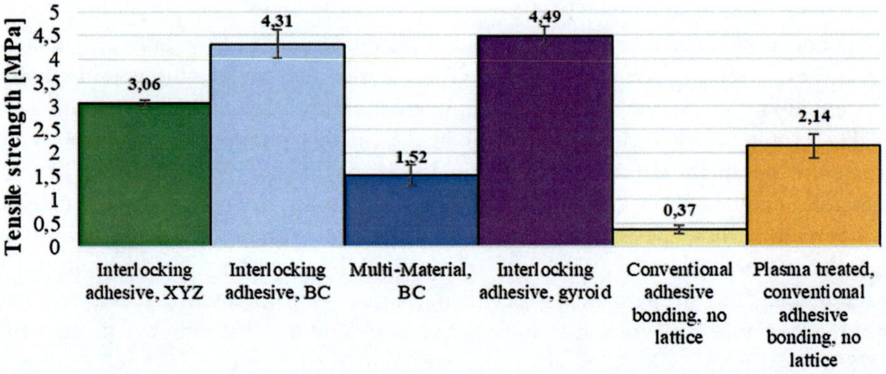

Fig. 7. Bonding AM polypropylene structures. Comparison of interlocking bonding to conventional means of adhesive bonding using Loctite HY 4070

In addition to the weakening of the lattice due to the semicircular cross-section of the horizontal struts, this weakening effect of the adhesive layer results in the structure with horizontal struts to be significantly weaker than the other tested geometries. This shows, that stiffness of the lattice is not the only metric by which these interlocking structures should be compared. Subsequently, since the strength of the lattice is already sufficiently high, further design measures to strengthen the lattice are unnecessary. However, since the failure is assumed to occur because of weakening of the adhesive at the very top of the lattice, decreasing the lattice density at the top is expected to further increase the strength. The gyroid structure shows slightly higher tensile strength than the BC structure for the same reason. Since the filling degree is significantly lower, the adhesive is suspected to be weakened less at the top of the lattice. Additionally, the weakening of the lattice through the small wall thickness of the structure is irrelevant, since the strength of the lattice still exceeds that of the adhesive. Nonetheless, all tested structures that utilize interlocking adhesive vastly exceeded the tensile strength of a plasma-pretreatment, demonstrating the viability of such structures for bonding low adhesive polymers.

Comparing bonding through interlocking adhesive to multi-material AM shows however, that the latter is significantly weaker even though the tensile strength of PETG is supposed to be higher than that of the used adhesive. Since failure exclusively occurred as delamination between PETG layers—a result that could not be reproduced in parts using only PETG—it is assumed that trace amounts of PP contaminated the PETG layers during manufacturing and reduced the auto-adhesion of the PETG. This means that even though multi-material AM despite its benefit in ease of handling cannot compete with structures that facilitate bonding by interlocking of an adhesive or even plasma treatment.

7 Conclusion and Future Prospects

This publication shows the design process along with the design freedom of lattice structures for various cases of application. Controlling stiffness, as well as orientation and density in different ways, is necessary to fully exploit the possibilities that lattice structures pose in designing novel products. Some design principles were established in order to maximize lattice stiffness, which were then applied to the use case of bonding by interlocking lattices. While it was shown that lattice stiffness is not the only mechanism that is affecting the tensile strength of such structures, it was shown that by applying these principles tensile strength of bonds with low adhesion polymers could be increased by more than 100% compared to plasma-pretreatment.

Funding

The research project IGF 21.138 N ("Design2Bond") by the Research Association on Welding and Allied Processes e. V. of DVS, Aachener Straße 172, 40223 Düsseldorf, Germany, is funded via AiF as part of the program for funding of Industrial Collective Research (IGF) by the Federal Ministry for Economic Affairs and Energy on the basis of a decision by the German Bundestag. Deutscher Verband für Schweißen [IGF 21.138 N].

Supported by:

Federal Ministry
for Economic Affairs
and Energy

iGF

on the basis of a decision
by the German Bundestag

References

1. Gebhardt, A., Kessler, J., Thurn, L. (eds.): 3D printing. Understanding additive manufacturing, 2nd edn. Hanser: Munich (2019). ISBN 978-1-56990-702-3
2. Gibson, I., Rosen, D.W., Stucker, B. (eds.): Additive manufacturing technologies. Rapid prototyping to direct digital manufacturing. Springer, New York (2010). ISBN 978-1-4419-1119-3
3. Vaezi, M., Chianrabutra, S., Mellor, B., Yang, S.: Multiple material additive manufacturing – Part 1: a review. Virtual and Physical Prototyping **8**, 19–50 (2013). https://doi.org/10.1080/17452759.2013.778175
4. Skylar-Scott, M.A., Mueller, J., Visser, C.W., Lewis, J.A.: Voxelated soft matter via multimaterial multinozzle 3D printing. Nature **575**, 330–335 (2019). https://doi.org/10.1038/s41586-019-1736-8
5. Gnanasekaran, K., et al.: 3D printing of CNT- and graphene-based conductive polymer nanocomposites by fused deposition modeling. Appl. Mater. Today **9**, 21–28 (2017). https://doi.org/10.1016/j.apmt.2017.04.003
6. Wu, J., et al.: Multi-shape active composites by 3D printing of digital shape memory polymers. Sci. Rep. **6**, 24224 (2016). https://doi.org/10.1038/srep24224
7. Rothe, S., Blech, C., Watschke, H., Vietor, T., Langer, S.C.: Material parameter identification for acoustic simulation of additively manufactured structures. Materials (Basel, Switzerland) (2020) 14. https://doi.org/10.3390/ma14010168
8. Ge, Q., Sakhaei, A.H., Lee, H., Dunn, C.K., Fang, N.X., Dunn, M.L.: Multimaterial 4D printing with tailorable shape memory polymers. Sci. Rep. **6**, 31110 (2016). https://doi.org/10.1038/srep31110
9. Watschke, H., Waalkes, L., Schumacher, C., Vietor, T.: Development of novel test specimens for characterization of multi-material parts manufactured by material extrusion. Appl. Sci. **8**, 1220 (2018). https://doi.org/10.3390/app8081220
10. Freund, R., Watschke, H., Heubach, J., Vietor, T.: Determination of influencing factors on interface strength of additively manufactured multi-material parts by material extrusion. Appl. Sci. **9**, 1782 (2019). https://doi.org/10.3390/app9091782
11. Nguyen, C.H.P., Kim, Y., Choi, Y.: Design for additive manufacturing of functionally graded lattice structures: A design method with process induced anisotropy consideration. Int. J. Precis. Eng. Manufacturing-Green Technol **8**(1), 29–45 (2019). https://doi.org/10.1007/s40684-019-00173-7
12. Vrána, R., Červinek, O., Maňas, P., Koutný, D., Paloušek, D.: Dynamic loading of lattice structure made by selective laser melting-numerical model with substitution of geometrical imperfections. Materials (Basel, Switzerland) (2018) 11. https://doi.org/10.3390/ma11112129

13. Abueidda, D.W., Jasiuk, I., Sobh, N.A.: Acoustic band gaps and elastic stiffness of PMMA cellular solids based on triply periodic minimal surfaces. Mater. Des. **145**, 20–27 (2018). https://doi.org/10.1016/j.matdes.2018.02.032
14. Freund, R., Koch, S., Watschke, H., Stammen, E., Vietor, T., Dilger, K.: Utilization of additively manufactured lattice structures for increasing adhesive bonding using material extrusion. J. Adhes. (2021), 1–22. https://doi.org/10.1080/00218464.2021.1983431
15. Grasshopper 3D. https://www.grasshopper3d.com/. Accessed 17 Febr 2022
16. ECR.labs. Dendro. https://www.ecrlabs.com/dendro/. Accessed 17 Febr 2022
17. Deshpande, V.S., Ashby, M.F., Fleck, N.A.: Foam topology: bending versus stretching dominated architectures. Acta Mater. **49**, 1035–1040 (2001). https://doi.org/10.1016/S1359-6454(00)00379-7
18. Deshpande, V.S., Fleck, N.A., Ashby, M.F.: Effective properties of the octet-truss lattice material. J. Mech. Phys. Solids **49**, 1747–1769 (2001). https://doi.org/10.1016/S0022-5096(01)00010-2
19. Ultimaker. Ultimaker PP Technical Data Sheet. https://support.ultimaker.com/hc/en-us/articles/360011963040-Ultimaker-PP-TDS
20. Henkel AG & Co. KGaA. Technisches Datenblatt - LOCTITE® HY 4070™. http://tds.henkel.com/tds5/Studio/ShowPDF/243%20NEW-EN?pid=4070&format=MTR&subformat=REAC&language=DE&plant=WERCS. Accessed 17 Febr 2022

Factories of the Future

Assisted Facility Layout Planning for Sustainable Automotive Assembly

Marian Süße[1](✉), Antje Ahrens[1], Valentin Richter-Trummer[1], and Steffen Ihlenfeldt[1,2]

[1] Fraunhofer Institute for Machine Tools and Forming Technology IWU, Reichenhainer Straße 88, 09126, Chemnitz, Germany
{marian.suesse,antje.ahrens,valentin.richter-trummer, buero.ihlenfeldt}@iwu.fraunhofer.de

[2] Institute of Mechatronic Engineering, Technische Universität Dresden, Wienerstr. 12, 28359, Dresden, Germany

Abstract. Decisions in factory layout planning can be considered multi-dimensional and complex since they need to cope with numerous partially conflicting boundary conditions and objectives. However, they do have a significant impact on long-term efficiency and flexibility. Due to rising needs in this area, an assisted solution for optimizing factory layout planning is required. Generative Design (GD) is a summarizing term for iterative, mostly nature-analogue approaches that support an efficient analysis of large design spaces, allowing to effortlessly achieve mathematically optimized solutions not usually achievable by traditional methods. Although there have been decades of research on the underlying principles, generative planning of spatial arrangements for manufacturing facilities still lacks behind its potential. Therefore, the proposed paper will begin with a structured overview of terminology and different factory planning requirements, followed by possible mathematical approaches for facility layout planning problems (FLP) in manufacturing. Special attention is drawn to sustainability aspects, defining the requirements to be considered in an automated design and including empirical knowledge in complex scenarios. The paper finishes with the derivation of identified future research areas.

Keywords: Factory of the future · Design and simulation · Generative design · Mathematical optimization · Sustainability

1 Introduction

The factory planning process may be understood as a process for system development that covers the overall production facility and its internal sub-systems. Since it is described and structured in numerous ways, several summarizing approaches and overviews were developed [1]. To generate a common baseline process, they were

K. Dröder and T. Vietor (Eds.): *Future Automotive Production Conference 2022, Zukunftstechnologien für den multifunktionalen Leichtbau*, pp. 173–188, 2023.
https://doi.org/10.1007/978-3-658-39928-3_13

transferred into the standardized guideline of the German society of engineers VDI 5200 [2]. In general, the procedure is divided into seven planning phases, which are in turn subdivided into further steps. The steps required for structural and layout planning are in phases 3 (concept planning) and phase 4 (detailed planning). The factory layout as a visual representation of the system elements is one primary result of factory planning. Thus, layout planning and factory planning may even be applied interchangeably in some superficial descriptions.

However, it is evident that the planning and design of factory layouts is a multi-dimensional and complex process that involves many stakeholders. General targets related to productivity and logistics need to be aligned with various restrictions and further demands (e.g., from construction, energy and media supply, or technical building supply). Moreover, once it is fixed and implemented, it has long-term effects on the efficiency and changeability of factory systems. At the same time, the increasing popularity of changeability is visible, for instance with the advent of *Reconfigurable Manufacturing Systems* (RMS). In such cases, it is a general approach to build up an infrastructure and select the appropriate flexible production elements that allow easy adaption as a response to altering market requirements or circumstances.

The layout design process still involves mainly experience-based decision processes and iterative discussions, often ignoring progress in mathematical optimization and available software tools. Therefore, more sophisticated and computer-based planning procedures may support an improved outcome for static layout design problems and increased speed for often recurring layout adaptions.

The following section provides a short classification starting with relevant terms related to the factory planning domain, specifying definitions for using *Generative Design* (GD) in technology, and providing the base for the overview of distinct corresponding algorithms. Subsequently, special requirements concerning automotive assembly design and sustainability constraints are carved out. The following section considers those, providing the authors' future approaches for tackling layout planning in environments with ever-increasing complexity and changing requirements. Special attention is given to the efficient integration of empirical knowledge based on an agile approach allied to iterative GD. An adaption of reinforcement learning then supports this as a promising solution approach, which introduces further challenges but is not yet state of the art for these types of problem sets. The paper ends with concluding remarks on challenges and requirements and provides a quick outlook on the next required research steps.

2 Classification and State of the Art

2.1 Terminology and Phases of Factory Layout Planning

The actual layout design process occurs in *concept planning* and *detailed planning* (see e.g. [2]).The result of the concept planning phase is an evaluated preferred variant of the real layout, which includes the dimensioned factory elements, material flow concept, and information and communication concept. Additionally, the associated preliminary building design is considered part of the concept stage. In the detailed planning phase, effects on the layout are limited to the first step of detailed planning. This step is characterized by an increase in the level of detail of operating equipment, logistics facilities, and all other factory system elements to form the basis for preparing specifications and tender documents. Accordingly, the specification of the equipment's orientation and media supply also occurs during concept and detailed planning. Typically, the planning phases are conducted sequentially and iteratively, involving extensive project management and coordination effort. Each phase comprises a set of decisions required to specify the factory system's dimension, shape, arrangement, connections, and elements. The applied decision-making strategies vary from qualitative selection processes to analytical and mathematical optimization methods. In most cases, the project outcome still depends on the discussion of the project teams and their experience. Figure 1 illustrates the basic planning phases and their inherent approaches.

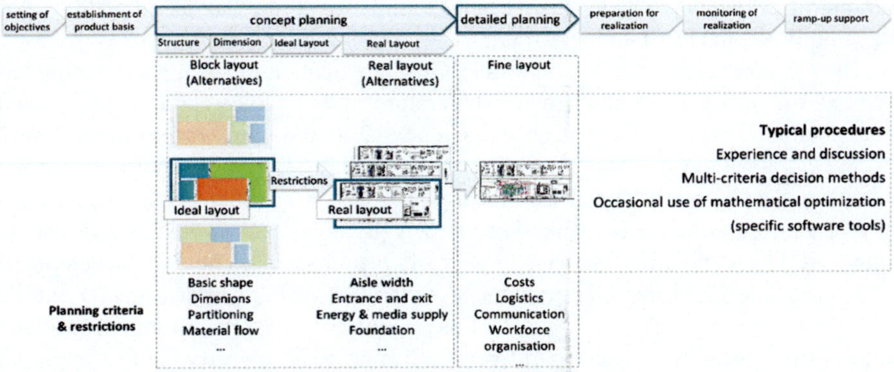

Fig. 1. Illustration of planning phases according to VDI 5200 and established decision procedures

An already withdrawn standardization with focus on material flow planning has been made in the VDI guideline 2498. Here, slightly deviating from the already indicated VDI 5200, the steps were divided into rough, ideal, actual, and detailed planning. The co-dependencies of material flow and layout planning are emphasized accordingly. There were also frequent references to other guidelines and links to energy and media supply. However, both VDI 5200 and VDI 2498 show that influencing variables from an ecological or energy perspective are only included in later planning stages of real or detailed planning.

Due to changing external factors in digitalization, software, and market requirements, constant request for adapting the sequential process is visible. On the one hand, this leads to adapting the project setup itself, for example, introducing agile project management philosophies [3, 4]. On the other hand, it raises demand to establish processes, roles, and tools of *Building Information Modelling* (BIM) as a major digitalization initiative in factory planning in the built environment [5]. Hence, factory planning is constantly confronted with adaptation requests to cope with increasing complexity. These developments interfere with the tools and approaches that may become more relevant in factory layout planning.

2.2 Specification of Generative Design as a Planning Principle

There may be a variety of available research for software and procedures, which could enhance quantitative decision-making for layout planning problems. GD is an increasingly popular term in technical development and construction, which is characterized by applying nature-inspired evolutionary principles to optimization tasks to generate efficient and creative design solutions. Thus, it should be highlighted that the application of GD principles does not necessarily aim to optimize technical or architectural systems. Instead, it supports the exploration of large design spaces that could not be analyzed manually by any human designer, architect, or factory planner. Therefore, the design based on generative principles may also relate to the so-called *Optioneering* (see, e.g. [5, 6]) that alludes to the systematic design process based on a large option set. However, GD may be seen as a subtype of automated layout design and optimization approaches, which relates to the broadly investigated Facility Layout Planning Problem (FLP) and its variants.

The relation between these, especially GD's application in the manufacturing environment, was investigated in a structured literature study by Süße and Putz [7]. It should be highlighted that GD was put on the same level as the terms *computational design* and *evolutionary design*. The analysis has shown that GD is well known in topology optimization and manufacturing and receives increasing attention in the built environment. Therefore, software tools like Spacemaker [8] are already well known and applied among architects and urban planners or the early planning stages of urban development.

A further result of this literature study is the identified gap related to GD in different layout planning problems. Thus, GD has mainly been applied in more static office buildings or exceptional cases apart from the business and industry context (e.g., planning refugee camps [9]). Therefore, the characteristics of contemporary manufacturing facilities (like resource and energy efficiency or changeability) require further development. So, a framework for integrating the conventional sequential planning process and the description of a GD model was introduced. A major aspect for future research is seen in the frontloading of detailed layout aspects to improve the specificity of generated layout variants. Furthermore, the integration of energy and ecology-related criteria is proposed.

With a comparable aim, Burggräf et al. [10] conducted a systematic literature review on the fields of action for automated layout design in factory planning. Thus, six specific areas for automating the design process of factory layouts were identified. These are briefly described in the following:

- The characteristics of the FLP as a general problem domain in Operations Research (OR) and optimization; Concerning this, systematic taxonomies and problem subtypes are described by Hosseini et al. [11].
- Multicriteria optimization, which alludes to the fact that the layout planning process needs to cover multiple target criteria and therefore requires appropriate decision methods
- The layout evaluation and selection especially come into play when multiple design alternatives reach the same result (e.g., Pareto optimization)
- Heuristics were also available in software applications, although there is much room for improvement regarding automation and multicriteria optimization

- Metaheuristics are well known for a broad application range in combinatorial problems; Besides local search strategies, they also cover nature-analog principles like evolutionary and genetic algorithms or population-based procedures
- The role of human planners, which relates to the model formulation and input data definition

One major aspect of future research is the interference of the algorithms, human planners, and required criteria selection, including the weight definition. As already indicated, the application of automated or GD processes does not replace human cognition and creativity. Especially the necessary formulation of the underlying optimization model requires mathematical knowledge and competencies in factory planning, especially when qualitative criteria should be integrated. Moreover, the design and development of an integrated planning system related to the BIM environment and tool landscape is proposed. In general, a closer collaboration between factory planners and software engineers is seen as a future necessity.

The terms and research areas mentioned above for automated layout design show interdependence and redundancies. Therefore, they may not be investigated separately. Moreover, most automated facility layout planning approaches refer to early planning phases and reduced decision criteria to handle complexity and improve calculation performance. Vice versa, the application of automated or GD-based layout planning becomes more challenging when fine layout aspects (e.g., assembly line planning), infrastructure restrictions, and sustainability requirements need to be covered.

2.3 Mathematical Formulations and Optimization

As facility layout planning has been a research area for decades, lots of work was already performed on finding problem formulations and approaching the automated and optimized design. Firstly, it is crucial to notice that the problem is NP-hard [12]. This definition from computational complexity theory describes that the time needed to solve a task is not in a polynomial relation to the task's input size. In the case of this work, with growing input size—e.g., the number of items to be placed—the time to solve the FLP to optimality rises to an unreasonable level. Hence, several different directions for categorizing and formulating such problems were found, and approximation methods were developed.

The review by Kusiak and Heragu [12], dating back to 1987, already includes five different modeling concepts—*quadratic assignment, quadratic set covering, linear- or mixed-integer programming,* and *graph-theoretic.* They also present many algorithmic approaches that are partially still vivid in the more recent research. This first glimpse at modeling is an indicator that numerous viewing directions and options for categorizing exist, as well as solving opportunities.

The analysis of Anjos et al. [13] describes FLPs with facilities of different sizing and categorizes the approaches according to the way the facilities are laid out– in rows (1D), on a plane with unequal areas (2D), or on multiple floors (3D). Various concrete formulations for these different conditions are presented, mainly focusing on *mixed-integer linear, semidefinite,* and *nonlinear optimization,* highlighting the characteristics of the mathematical formulation. They also link to several approximation techniques like branch-and-bound, genetic algorithms or two-stage approaches.

Drira et al. [14] start their review with the definition of FLP, which is not consistent in literature, and compose a tree representation considering the categories workshop characteristics, addressed problem, and solution approaches. To give an impression of the factors within these categories, here are some examples:

- the handling system, also corresponding to the aforementioned 1, 2, 3D system boundaries,
- the dynamic behavior (i.e., whether the layout should be flexible over time),
- the chosen formulation, e.g. as discrete, continuous, or fuzzy.

They also point out that realistic FLP has multiple objectives contrary to the assumptions in most research up to their review.

In another recent review, Pérez-Gonsende et al. [15] analyzed 232 reviewed articles regarding the problem type, approach, applied planning phase, facility characteristics, and the actual layout generation principle. As an outcome, they state that mainly discrete quadratic programming models for equally sized departments as well as continuous linear and nonlinear mixed-integer programming models for different sized departments are applied. Moreover, specific approaches for the integration of expert knowledge were found. In general, it is stated that the most frequent solution algorithms were metaheuristics, with about 80% being genetic algorithms.

As genetic algorithms are a prominent option for NP-hard problems, it is no surprise that they are prominent for FLPs. Compared to genetic algorithms, other methods of the AI spectrum that are a constantly growing field in recent years, it is rather counter-intuitive that it is not that present in the FLP research. In the review by Burggräf et al. [16], another overview of different categorizing methods is presented and shows the distribution of publications depending on these. They provide a comprehensive overview of the activities using machine learning. In the resume, ML emerges in the FLP field but is rarely used for solving the actual problem. Possible reasons are that it is difficult to define the learning task and to generate labelled data for supervised learning, which is not biased by the respective expert. The subsequent Table 1 provides a qualitative overview on the previously mentioned literature reviews.

Table 1. Comparison of literature reviews on facility layout planning problems

Authors	Aspect			
	Overview on model formalization	Evolutionary solution approaches	ML-based solution approaches	Sustainability-related criteria
Kusiak and Heragu [12]	●	◑	◌	◌
Anjos et al. [13]	●	◑	◌	◌
Drira et al. [14]	●	●	◑	◌
Pérez-Gonsende et al. [15]	●	●	◑	◑
Burggräf et al. [16]	●	◑	●	◌

◌ not included; ◑ partial reference; ● extensive consideration

As a key result of the meta-level literature study it can be stated that mathematical formulations for the optimization of facility layouts have been documented extensively. Besides that, evolutionary approaches are very well investigated in the field of NP-hard problems. At the same time, it becomes obvious that ML-based solution strategies reoccur occasionally but relevance, especially in real application scenarios, is still reduced. Moreover, it becomes clearly visible that sustainability-related optimization goals and boundaries are so far of marginal importance in the optimization approaches. With reference to this, requirements from sustainability-related aspects are introduced in Sect. 3.

3 Elaboration of Specific Requirements

3.1 Planning of Automotive Assembly Lines

Automotive assembly should be separated into at least two distinct areas. Firstly, *body-in-white assembly* is usually performed with the highest automation degree possible. Secondly, *final assembly* with a significant amount of manual work, assisted by technical solutions whenever possible. This distinction is mainly due to available automation capabilities and might change in the future. While automotive assembly is not entirely different from other automated assembly lines, some specificities have to be considered for the automated planning of automotive assembly lines.

Besides low-cost, high rate production, a primary requirement of the automotive industry is the highest possible process *robustness*. This robustness includes technological aspects such as joining technologies, material choice, and tight single part tolerances for achieving the highest quality assembled parts. However, additionally, it includes robustness in terms of resilience against external factors or internal effects, such as machine defects. Different strategies can be followed to achieve resilience: *duplication* and *parallelization* of assembly lines and more recently investigated island-based or *modular assembly* cells (see, e.g. [17, 18]). Such design decisions also significantly impact an optimized intralogistics solution, which has to be integrated into automated factory design. A further aspect with increasing importance is the adaptability of the factory layout to new car models, partially even intending to produce all car models in mixed sequence. Since car models are redesigned in shorter periods than assembly stations could technically last, at least two different approaches can be chosen for optimizing production sustainability. Firstly, platform-based modularity in the car design allows reusing assembly stations at least partially. Secondly, adaptable jigs and tools (see, e.g. [19]) can significantly increase the reuse of existing production systems. Adaptability can be achieved either by a rigid factory layout with adaptable tools only or by allowing complete adaptability in the factory layout. Another typical requirement is the ability to ensure consistent assembly quality. Usually, product design already includes easily measurable characteristics for this purpose. Two distinct options should be considered—either in-line measurement can be used, or a separate measurement room, including the required intralogistics system, must be considered. In this case, traceability, both of the product and the production system, is also essential since it allows recognizing complex cause-effect relationships, which can influence assembly quality.

3.2 Sustainability as an Overarching Paradigm

As a summarizing term for mitigating external risks and hazards caused by goods manufacturing, sustainability is broadly applied for strategic agenda setting in the industry. Numerous definitions and interpretations exist. One well-known definition refers to the *Triple Bottom Line* [20], which comprises ecological, environmental, and social impacts. A more differentiated set of efforts is provided by the United Nation's *Sustainable Development Goals* (SDGs) [21], defined in 2015. The first one seems to focus more on entrepreneurial behaviour and business context; the latter refers to the overall future development of peaceful and prosperous societies. However, both interpretations describe the concept on a general level, which still requires specification and the development of target-oriented metrics for decisions in manufacturing and especially in factory planning.

General guidelines and principles for reducing the environmental impact of factories are thus well known [22, 23]. For instance, maximizing the use of natural light is a typical demand for the facility design to reduce electrical energy consumption. Hence, tools for lighting analysis have already been integrated into available software solutions allowing the investigation of building designs, e.g., in the established BIM software Autodesk Revit [24]. Especially in the case of assembly processes where minimum light intensities are required for specific operations.

Also, the minimization of land use while still keeping the ability to modify the factory layout in the future is a typical proposal. Whenever possible, grouping machines that need the same type of energy may be an intuitive approach to reduce energy losses. Additionally, more complex design aspects such as an optimized natural airflow can improve sustainability.

However, while product and production planning and building architecture probably have the highest impact on environmental sustainability, some layout-related aspects may be underestimated in industrial practice. A *framework for GD-based* layout planning enhanced by ecological criteria has previously been developed and published by Süße et al. [25]. As a baseline structure, the selected environmental impact categories for the layout planning framework were aligned with the selected mid-point categories of the Organization Environmental Footprint, defined by the European Joint Research Center [26]. Those were combined with pre-existing taxonomies of FLP goals and constraints. Figure 2 depicts the relation between classical criteria and further aspects that affect the environmental performance of a designed factory layout.

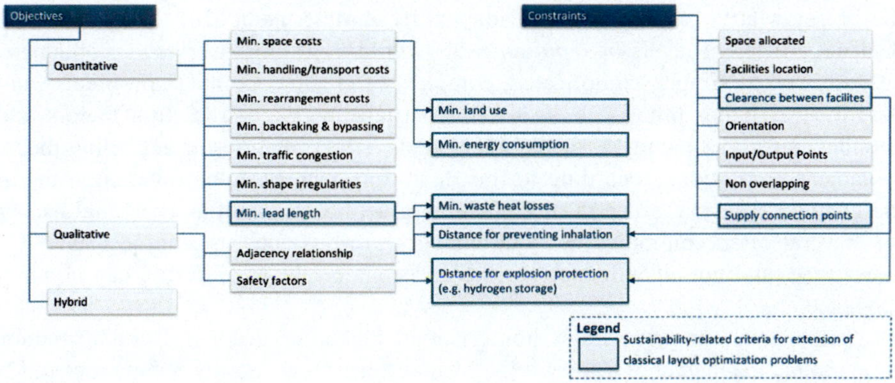

Fig. 2. Extended optimization goals and restrictions for sustainability improvements in layout optimization

It becomes evident that some established optimization goals already contribute to reducing environmental impacts. Primarily referring to the quantitative aspects, the monetary target of minimized space costs supports reducing land use and energy consumption in general. Moreover, reducing transport costs also reduces energy consumption and emissions of transport elements. The lead length (e.g., in energy and media supply) and distances between heat sources and heat sinks are further layout-related criteria that could be minimized to improve overall factory system sustainability.

Moreover, the mathematical formulation can be augmented with additional constraints. For instance, connection points for energy supply need to be considered, and clearance between facilities (i.e., minimum distances) is required to protect humans from chemical emissions or pollutants. The same accounts for renewable energy sources like hydrogen-based Power-to-Gas systems, making safety distances between gas tanks and machining operations necessary.

However, the addition of criteria expands the number of dimensions to be optimized and may lead to target conflicts. For example, minimizing transport costs and waste heat losses reduce distances between different facilities, whereas safety considerations may require larger distances between areas. An overall complex layout generation problem can therefore be expected. Besides that, the appropriate quantification factors and mathematical models will have to be specified to support an integrated sustainability assessment in the layout design phase.

4 Future Concepts

4.1 Integration of Empirical Knowledge

According to Stacey and Mowles [27], problems can be categorized by the *level of agreement among stakeholders* regarding a decision to be taken and the certainty regarding decisions to reach a particular objective. This relation is usually known as the Stacey Matrix, which has been adapted by Zimmermann [28] to explain how to deal with different levels of *requirements.* According to this overview, problems can be described as simple, complicated, complex, or chaotic. Complex problems can be described as being influenced by several partially even contradicting factors while simultaneously not having a visible linear cause-effect relationship regarding the way to achieve a solution. According to this definition, factory layout planning falls into the *complex* category, since during planning greenfield as well as brownfield scenarios, several often contradicting requirements from different stakeholders have to be considered, and not all influences of the decisions on the final design can be readily predicted.

Furthermore, empirical know-how has to be taken into account. In such situations, agile project management frameworks, such as Scrum, are usually recommended [29]. While Scrum is usually applied to software development, the concept can also be used in various other challenges, among others, for assembly or layout planning. For the application described in the present article, the artifact *product increment* is probably the most important since it leads to the requirement that factory planning should not occur in a fully automated manner based on requirements fully defined right from the beginning. Instead, an *iterative approach* allows stakeholders to analyze development increments and adapt requirements along with the development according to their empirical knowledge to reach the most promising factory layout.

A fully automated design solution would most likely identify the global optimum solution of such a problem, if the problem can be fully specified, and optimization criteria are weighted according to a general agreement between all stakeholders before starting calculations. Since, in reality, neither full specification nor a perfect weighting can be guaranteed for such complex projects before design starts, it is illusory to assume that a global optimum can be reached based on the initial definition. Therefore, it can be accepted that incremental development will allow reaching the best possible solution for a given set of stakeholders and goals.

The Scrum framework is based on three roles (product owner, developer, scrum master), three artifacts (product backlog, sprint backlog, product increment), and five events (sprint, sprint planning, daily Scrum, sprint review, sprint retrospective). The goal of the automated factory layout planning system is to minimize manual activities in this framework for the specified problem. Regarding the roles, mainly the *developer* is replaced by the envisaged algorithms. The scrum master can be replaced or at least strongly supported by a *self-explaining GUI,* leading the user to the most helpful workflow. The product owner represents all involved stakeholders with their specific requirements for the final factory layout. This role is assisted with a suitable GUI, which joins and visualizes all requirements for the different parties, enhancing transparency and thus co-development. The *product backlog* is linked to the visualization of all requirements and is an essential aspect of the GUI.

In contrast, the sprint backlog is of lower importance for automated *product increment* calculations due to the virtually unlimited speed of the envisaged algorithms. This calculation speed also reduces the need for any sprint planning and daily Scrum, and also the actual sprint duration can be virtually zero. The sprint review is again facilitated by a suitable GUI, allowing all parties to analyze the outcome of the calculated factory layout and comment on it in an interactive and concurrent environment.

Retrospective aspects would be mainly performed in the development phase of the envisaged software and not in a large amount in the final usage phase of the software, which will also be developed following the Scrum framework. Therefore, the goal of the developed algorithms is to provide almost immediate *product* (i.e. factory layout–)*increments*. These provide direct and transparent feedback to the stakeholders involved in factory layout definition, especially regarding the combination of their requirements, including an automated evaluation of pre-defined key figures required for objective comparison and further decisions. Therefore, the presented approach facilitates *transparency* and verifiability to allow adaptations to the design based on empirical knowledge.

4.2 Automated Planning by Enhancing Generative Design

As pointed out before, generative design gives way to the exploration of large design spaces. Therefore, a methodology needs to be implemented for calculating designs and rules on how to explore/improve further. Work in this field was already done using nature-inspired algorithms such as genetic algorithms. While these are promising paths to follow, a challenge could be to integrate empirical knowledge as described in the previous paragraph. Because this would require a recurring refactoring of the problem setup, it could also require the recurring refactoring of the genetic algorithm's parameter like the number of iterations or population size.

Therefore, another idea for implementing GD could be using *reinforcement learning* (RL), which did receive almost no attention in the past for FLPs. RL, in general, is a *machine learning* (ML) approach of 'trial & error' with a variety of applications in several research areas (see, e.g. [30]) and very decent success. The main idea is to rate a current state with a reward function and deviate a transition to another state with a better reward. In the case of FLP, the states would be different layouts with defined origin and dimensions for every facility. A rough sketch of this state-action-space could be formulated in the following way:

- As preliminary, the desired result is a 2D-floor plan of a facility with origin in (0,0) and area in $\mathbb{R}^2_{\geq 0}$ only, the departments are rectangular and aligned to the x-y-axes
- Let $D = \{d_1, \ldots, d_n\}$ be the set of departments to be placed with $|D| = n \in \mathbb{N}$ the number of departments
- a state for one facility d_i is a 4-dimensional array $\{x_i, y_i, l_i, w_i\}$ with
  - $\{x_i, y_i\} \in \mathbb{R}^2_{\geq 0}$ being the coordinates of the origin/lower-left corner of the departments' rectangle,
  - $l_i \in \mathbb{R}_{>0}$ the dimension in the x direction, $w_i \in \mathbb{R}_{>0}$ the dimension in the y direction
- An overall state s would be a matrix M_s of dimensions $(n \times 4)$, with every row corresponding to one department
- Action a would then be defined as the transition between two state matrices $M_s, M_{s'}$, or in other words, as the variation of the departments' origin and dimensions

The goal or reward/punishment function combines and weights all the layout requirements defining the learning algorithm's development direction. Firstly, if the locations defined in one state would violate any external constraints, e.g., the size and accessibility to necessary supplies, they would be punished in the rating. As we talk about mandatory constraints, the punishment must be defined in a way the algorithm tries to avoid these in later periods.

As an example, most probably every department d_i has a minimum size (s_i^x, s_i^y). Therefore, in the goal function, we can add the term $P \cdot (l_i - s_i^x)$ as well as $P \cdot (w_i - s_i^y)$ with P being the weight/scale of the punishment. It should be chosen as an enormous, positive value while the numbers in brackets are negative, but either zero or negative (but not necessarily as large) if they are positive. Otherwise, the algorithm would learn to oversize as much as possible. As a side remark, we can conclude that the weights are not obligatory numbers only but can be (plain) functions as well.

Secondly, optimization depending on different objectives can be applied almost arbitrarily through the reward function if all constraints are fulfilled. If, for example, no outer limitation on the dimension exists, one could aim for a relatively compact layout, which can be quantified by the area it consumes. Qualifying this further, more specifications for the layout need to be defined, which are illustrated in Fig. 3. There, three departments are shown, and the minimal rectangular including them is the facility.

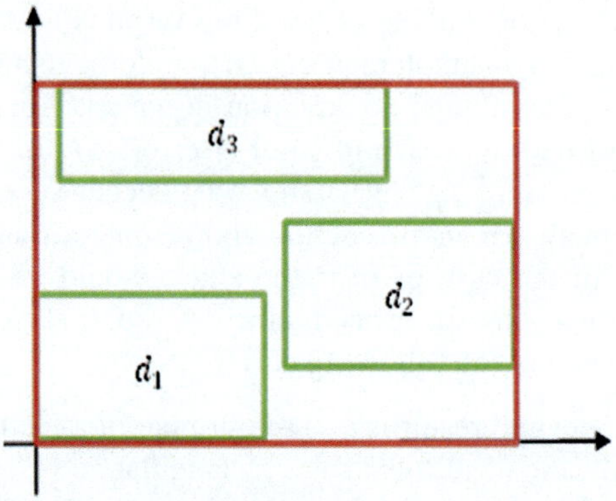

Fig. 3. Possible floor plan with three departments and overall facility boundary

With that being said, the area of the facility is

$$A = (x_{i^*} + l_{i^*}) \cdot \left(y_{j^*} + w_{j^*}\right) \text{ with}$$
$$i^* = argmax_{i=1,\ldots,n}\{x_i + l_i\}, \text{ and}$$
$$j^* = argmax_{j=1,\ldots,n}\left\{y_j + w_j\right\}.$$

The reward for this area could simply be the negative $-A$ or could also be a product with a weight. This depends on the other factors in the reward function and is to be explored in future work.

More advanced requirements would consider, for example, the distance to supply fixed points. These can be formulated as a boundary like the minimum size if an upper limit on the distance from a department to such a point exists. Alternatively, or additionally, it can be formulated as the area one if these distances should be minimized. Then the weighting plays a vital role, connecting to the result from the sprint result.

The general advantage of using RL is that it is unnecessary to define any strategy for solving the task. Instead, the algorithm develops a strategy based on the boundaries and objectives. It is also possible to integrate expert knowledge in the RL learning process, as in interactive or inverse RL [31, 32].

Additionally, ongoing research targets the issue of adapting to changing environments so that varying specifications or weighting can be integrated as well [33]. On the downside, RL is generally computationally intensive. As the state space in FLP is almost infinite (since dimensions are continuous), the question arises of how it is possible to calculate—and re-calculate—in a reasonable amount of time. As the last-mentioned supplementary ideas are still active research fields, the quality of the results needs to be confirmed.

5 Conclusions

Factory planning is an increasingly complex task due to ever-increasing requirements and shorter product and, therefore, production system life cycles. Not only do requirements refer to higher expected precision, but also the type of requirements is evolving, especially in terms of environmental and societal impact. This variety of requirements leads to the need to solve very complex problems in a short time, using the minimum amount of resources possible. Therefore, developing automated or assisted planning systems is logical and required.

As an active research area for decades, lots of effort has been put into structuring and solving this challenge. Activities include classifying the planning steps and finding manual as well as automated solutions. However, particular attention should be paid to the possibility of applying generative design as a solution principle since it can further expand the possibilities of mathematical modeling.

Looking at current and future requirements in manufacturing, especially in the light of automotive assembly, it is clear that the number of possibly conflicting objectives increased over the past years and will continue to do so in the future. The ongoing strive for more flexibility creates new demands for dynamic layout changes. Therefore, future facility changes and variations must be considered in the initial planning phase.

Moreover, the rising *focus on sustainability* also raises entirely additional perceptions and challenges for factory planning, e.g., by taking the reduction of waste heat losses into account. Since such requirements will not necessarily always be in line with the most economical solution, methods need to be developed or adapted to handle conflicting goals throughout the planning phase transparently. It is proposed to handle such complex scenarios by applying agile management methods and integrating an adapted version of such frameworks into the automated planning approach. With this approach, every iteration step can be analyzed and openly discussed, adapting the methods and goals to best match stakeholders' requests without significantly increasing the planning effort. Generative design has been chosen as the most promising path for each iteration step, which can be executed with metaheuristic algorithms or reinforcement learning. Figure 4 summarizes this development.

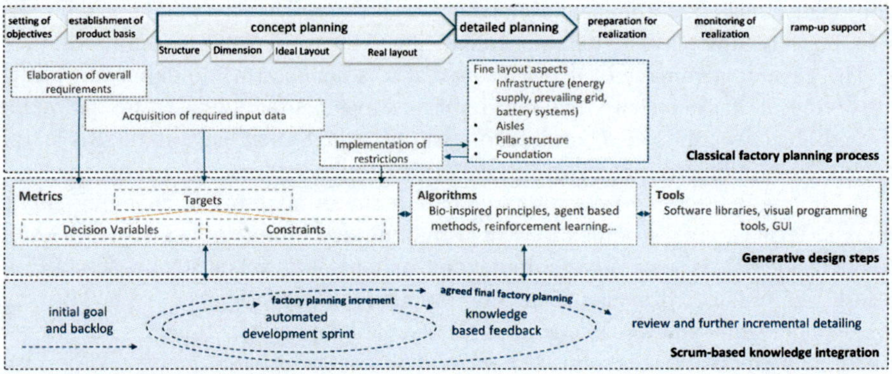

Fig. 4. Summarizing concept of the envisaged solution

The goal is to integrate the discussed topics into a single software solution, which can be tested with a real-world use case. Further research is required to increase applicability for project stakeholders in later project phases and multi-dimensional decision problems.

References

1. Michael Clauß: Methode zum Einsatz von Web 2.0-Werkzeugen in der Fabrikplanung (2013). https://monarch.qucosa.de/api/qucosa%3A19915/attachment/ATT-0/. Accessed 22 Feb 2022
2. VDI 5200 Blatt 1 Fabrikplanung—Planungsvorgehen, VDI-Gesellschaft Produktion und Logistik (Mar. 2011)
3. Augustin, H., Al-Shahmani, H., Dreßler, E., Krafzik, C., Liebler, C., Schmidt, P.: Gestaltung eines virtuellen Fabrikplanungsprozesses. ZWF **115**(10), 659–662 (2020). https://doi.org/10.1515/zwf-2020-1151003. Accessed 3 Feb 2022
4. Reinema, C., Pompe, A., Nyhuis, P.: Agiles Projektmanagement. ZWF **108**(3), 113–117 (2013). https://doi.org/10.3139/104.110903
5. Brunone, F., Cucuzza, M., Imperadori, M., Vanossi, A.: An innovative method for the management of the building process. In: Brunone, F., Cucuzza, M., Imperadori, M., Vanossi, A. (Eds.) Springer eBook collection, wood additive technologies: Application of active design optioneering, 1st ed., pp. 35–64. Springer, Cham (2021). Accessed 3 Mar 2022
6. Keshavarzi, M., Rahmani-Asl, M.: GenFloor: Interactive generative space layout system via encoded tree graphs. Front. Archit. Res. **10**(4), 771–786 (2021). https://doi.org/10.1016/j.foar.2021.07.003. Accessed 1 Feb 2022
7. Süße, M., Putz, M.: Generative design in factory layout planning. Procedia CIRP **99**, 9–14 (2021). https://doi.org/10.1016/j.procir.2021.03.002. Accessed 23 Mar 2022
8. Spacemaker A.S.: Spacemaker—AI architecture design | building information modelling. https://www.spacemakerai.com/. Accessed 4 Mar 2022
9. Elie, D., Kubicki, S.: Technologies in the planning of refugees' camps: A parametric participative framework for spatial camp planning. In: 2017 IEEE Canada International Humanitarian Technology Conference (IHTC), Toronto, ON, Canada, Jul. 2017, pp. 207–212. Accessed 22 Feb 2022

10. Burggräf, P., Adlon, T., Hahn, V., Schulz-Isenbeck, T.: Fields of action towards automated facility layout design and optimization in factory planning—A systematic literature review. CIRP J. Manuf. Sci. Technol. **35**, 864–871 (2021). https://doi.org/10.1016/j.cirpj.2021.09.013. Accessed 1 Mar 2022
11. Hosseini-Nasab, H., Fereidouni, S., Fatemi Ghomi, S.M.T., Fakhrzad, M.B.: Classification of facility layout problems: A review study. Int. J. Adv. Manuf. Technol. **94**(1–4), 957–977 (2017). https://doi.org/10.1007/s00170-017-0895-8
12. Kusiak, A., Heragu, S. S.: The facility layout problem. Eur. J. Oper. Res. **29**(3), 229–251 (1987). https://doi.org/10.1016/0377-2217(87)90238-4. Accessed 3 Dec 2021
13. Anjos, M.F., Vieira, M.V.: Mathematical optimization approaches for facility layout problems: The state-of-the-art and future research directions. Eur. J. Oper. Res. **261**(1), 1–16 (2017). https://doi.org/10.1016/j.ejor.2017.01.049. Accessed 1 Mar 2022
14. Drira, A., Pierreval, H., Hajri-Gabouj, S.: Facility layout problems: A survey. Ann. Rev. Cont. **31**(2), 255–267 (2007). https://doi.org/10.1016/j.arcontrol.2007.04.001. Accessed 11 Dec 2021
15. Pérez-Gosende, P., Mula, J., Díaz-Madroñero, M.: Facility layout planning. An extended literature review. Int. J. Prod. Res. **59**(12), 3777–3816 (2021). https://doi.org/10.1080/00207543.2021.1897176. Accessed 15 Mar 2022
16. Burggräf, P., Wagner, J., Heinbach, B.: Bibliometric study on the use of machine learning as resolution technique for facility layout problems. IEEE Access **9**, 22569–22586 (2021). https://doi.org/10.1109/ACCESS.2021.3054563
17. Burggräf, P., Dannapfel, M., Adlon, T., Kahmann, H., Schukat, E., Keens, J.: Capability-based assembly design: An approach for planning an agile assembly system in automotive industry. Procedia CIRP **93**, 1206–1211 (2020). https://doi.org/10.1016/j.procir.2020.03.079. Accessed 21 Feb 2022
18. Kern, W., Lämmermann, H., Bauernhansl, T.: An integrated logistics concept for a modular assembly system. Procedia Manufac. **11**, 957–964 (2017). https://doi.org/10.1016/j.promfg.2017.07.200. Accessed 22 Feb 2022
19. Neugebauer, R., Putz, M., Pfeifer, M., Todtermuschke, M.: Improving flexibility in car body assembly systems. CATS 2010: Responsive, costumer demand driven, adaptive assembly, pp. 115–120. Trondheim (2010). Accessed 11 Feb 2022
20. Elkington, J.: The triple bottom line. In: M. V. Russo (Ed.): Environmental management: Readings and cases, 2nd ed., pp. 49–66. SAGE, Los Angeles (2008). Accessed 15 Jan 2022
21. United Nations, THE 17 GOALS | Sustainable development. https://sdgs.un.org/goals. Accessed 4 Mar 2022
22. Müller, E., Engelmann, J., Löffler, T., Strauch, J.: Energieeffiziente Fabriken planen und betreiben. Springer Berlin Heidelberg, Berlin (2009)
23. Nielsen, L., et al.: Towards quantitative factory life cycle evaluation. Procedia CIRP **55**, 266–271 (2016). https://doi.org/10.1016/j.procir.2016.08.009
24. Autodesk Inc.: About lighting analysis | Revit 2021 | Autodesk knowledge network. https://knowledge.autodesk.com/support/revit/learnexplore/caas/CloudHelp/cloudhelp/2021/ENU/Revit-Analyze/files/GUID-1F9669B3-338B-436D-B850-3FA4BC84A300-htm.html. Accessed 4 Mar 2022
25. Süße, M., Ihlenfeldt, S., Putz, M.: Framework for increasing sustainability of factory systems by generative layout design. Procedia CIRP **105**, 345–350 (2022). https://doi.org/10.1016/j.procir.2022.02.057. Accessed 5 Apr 2022
26. Pelletier, N., Allacker, K., Manfredi, S., Chomkhamsri, K., Maia de Souza, D.: Organisation environmental footprint (OEF) guide. https://ec.europa.eu/environment/eussd/pdf/footprint/OEF%20Guide_final_July%202012_clean%20version.pdf. Accessed 19 Sep 2021

27. Stacey, R. D., Mowles, C.: Strategic management and organisational dynamics: The challenge of complexity to ways of thinking about organisations. Pearson Education, Harlow (2016). Accessed 28 Feb 2022
28. Zimmerman, B.: Ralph Stacey's agreement & certainty matrix. https://www.betterevaluation.org/en/resources/guide/ralph_staceys_agreement_and_certainty_matrix. Accessed 4 Mar 2022
29. Schwaber, K.: Agile project management with Scrum. Microsoft Press; Safari Books Online, Redmond (2004). https://learning.oreilly.com/library/view/-/9780735619937/?ar. Accessed 4 Mar 2022
30. neptune.ai: 10 real-life applications of reinforcement learning—neptune.ai. https://neptune.ai/blog/reinforcement-learning-applications. Accessed 4 Mar 2022
31. Arzate Cruz, C., Igarashi, T.: A survey on interactive reinforcement learning. In: Proceedings of the 2020 ACM designing interactive systems conference, Eindhoven, Netherlands, pp. 1195–1209 (2020). Accessed 11 Feb 2022
32. Arora, S., Doshi, P.: A survey of inverse reinforcement learning: Challenges, methods and progress. Artif. Intell. **297**, 103500 (2021). https://doi.org/10.1016/j.artint.2021.103500. Accessed 25 Feb 2022
33. Padakandla, S.: A survey of reinforcement learning algorithms for dynamically varying environments. ACM Comput. Surv. **54**(6), 1–25 (2021). https://doi.org/10.1145/3459991. Accessed 26 Feb 2022

Assembly and Through Life Services in the Context of Urban Cloud Manufacturing

Aydan Oguz(✉), Pinar Bilge, Arne Glodde, Sina Rahlfs, and Franz Dietrich

Institute for Machine Tools and Factory Management, Chair for Handling and Assembly Technology, Technical University Berlin, Pascalstraße 8-9, 10587 Berlin, Germany
{aydan.oguz,p.bilge,arne.glodde,s.rahlfs, f.dietrich}@tu-berlin.de

Abstract. Platform economy enables entirely new value creation processes with remarkable resistance to crises compared to centralized production systems. Acknowledged samples in commerce, travel accommodations, and creative media platforms are well known. Samples in production technology are, despite the potentials in platform economy, still rare. Its application in complex production systems must contribute to individualized product and process design and realization. Thus, research and development on structuring and utilization in a platform-supported production economy are required. We propose the framework "Urban Circular Cloud Assembly and Services (Urban CIRCLAS)", which is developed for granularization and order-specific ad-hoc new combination of operating resources, processes and workers. The Urban CIRCLAS framework contains handling technology, measurement technology, manual and automated assembly systems as well as a cloud-based IT infrastructure for demonstration of feasible circular production samples in an urban context. This paper presents the concept, potential and challenge of the Urban CIRCLAS framework. It highlights how users such as freelancers or small companies combine and upgrade their know-how and capabilities to realize innovative product and process designs, which meet established industry standards. This is elaborated by three scenarios with different requirements. Based on the outcomes, potential application and research areas enabled by the Urban CIRCLAS framework are discussed. As a result, important insights can be gained from the concept of the Urban CIRCLAS.

Keywords: Circular economy · Platform economy · Assembly system · Urban production

K. Dröder and T. Vietor (Eds.): *Future Automotive Production Conference 2022*, Zukunftstechnologien für den multifunktionalen Leichtbau, pp. 189–204, 2023.
https://doi.org/10.1007/978-3-658-39928-3_14

1 Introduction

The shift from cradle-to-grave to a circular requires the development of new concepts. Sustainable development is one of the main topics of daily news. Movements like Fridays for Future have started [1]. Europe responded to this change through achievements and regulations like the European Green Deal [2]. Therefore, the sustainability challenge must be met from the customer's point of view as well as from the point of view of legislation, while it forces companies to rethink their production. This paper aims to identify unexploited value creation patterns and provides a new flexible and easily reconfigurable framework to meet this challenge of sustainability and humanisation.

The goal of the proposed framework is order-specific ad-hoc combination of operating resources, processes and workers with handling, measurement and assembly technologies and methods. The following outline summarizes the structure of the paper. Section 2 identifies new approaches in assembly systems and evaluates the potential of urbanization in terms of sustainability as well as manufacturer and customer attractiveness. It also describes the principle of platform economy, outlines the need for research and development. Based on the findings, the Urban CIRCLAS framework is elaborated in Sect. 3. Section 4 exemplifies the usefulness of the new framework based on user scenarios. They illustrate the advances of the framework such as saving of resources, maximizing output per area.

2 Related Word and Scope of this Paper

The principle of circular economy becomes increasingly important for governments, customers and companies. However, production companies must face enormous challenges to satisfy regulative and customer requirements. The trend of production shifts from a mass production to mass customization [3]. Increasing product variants challenges facilities caused by higher production costs, which leads to investigate in new concepts [4]. The integration using opportunities such as digitization enables to address sustainability aspects, as waste from conventional mass production is reduced [5].

An increase of the urban population is expected in the 21st century [6]. This implies a change to urban production, which offers a close-to-market option as well as counteracts the increasing environmental demands [6, 7]. Urban processes are mostly next to the use stage of the products. Thus, service processes such as maintenance, repair and overhaul (MRO) can be realized. In order to comply with the principle of circular economy, the cycle must be closed by appropriate end-of-life (EoL) stages. EoL aims a second-life, which means that process of and design for disassembly becomes increasingly important.

Companies are therefore committed to meet the trend of mass customization as well as to increasingly integrate the EoL stages into their production. The connection to multiple stakeholders is essential but also a legislative and organizational challenge. Thus, the platform economy regulates and simplifies such cooperation. In anticipation of this development, research is conducted in a wide range of production areas [8]. The following subsections explain the current findings as well as applications with regard to changes in market requirements and increasing urbanization.

2.1 Circular Economy

With the advent of environmental regulations around the world, the term circular economy has become increasingly important [9]. The term circular economy (CE) describes a paradigm, that consumer goods are used for as long as possible. To achieve this, it has to be possible to repair the product and disassemble it at the end of its life cycle. At best, it is enabled to reuse the individual parts in other systems or areas. The recycling of waste is the last step taken to ensure at least a recovery of raw materials [9]. To meet this goal, companies must adapt their business model to the CE paradigm.

Small and medium-sized enterprises (SME) account for 90% of the world's business and 50–60% of the workforce is employed in SME [10]. However, especially for SME the adaption to a CE paradigm appears to be much more challenging than for larger companies [10]. The factors that influence a successful integration of CE are divided into external and internal factors. External factors include public policy, market conditions, the ability to develop new technologies and potential stakeholders. Internal factors describe the company's own resources, capabilities, and competencies [10]. The CE paradigm is a main part of the Urban CIRCLAS concept development. Processes must be identified which profitably implement the circular economy paradigm. An urban production system enables advantageous market conditions in terms of CE paradigm and the market requirements.

2.2 Urban Production

One of the most influential factors at the time of Industry 4.0 is the increasing urbanization. Considering the European area, the share of urban population will be 80% of the total population in 2050. Thus, a change of product requirements and a demographic change in production locations are essential [6].

The term "urban production" describes the transformation of production to the city. It is the interaction of urban populations and the companies. Accordingly, urban production also demands production and resource-saving processes to minimize noise and environmental emissions. On the one hand, it offers the possibility of implementing the principle of individualized products directly in the city for the customer, and on the other hand, a workplace in the immediate area of the own living space has a higher attractiveness for potential workers [6, 11]. Based on this, urban production requires high productivity in limited spaces, which are more expensive in big cities than uninhabited areas. These requirements necessitate an adaption of production. Due to space limitations, productivity must be increased by other measures than economies of scale. This provides a new option for decentralized production in cities, especially for SME [10]. Thus, production in small, decentralized units offers potential for economic impacts and climate-friendly effects, especially in urban environments [6].

The proposed Urban CIRCLAS concept supports decentralized production in urban areas. It focuses on high productivity on a small area and uses the advantages of an urban environment to reduce logistical effort. The Urban CIRCLAS concept also reduces efforts for various production processes. Thus, different assembly systems are carried out in the following.

2.3 Assembly Systems and Factories

Individualized design demands as well as the avoidance of overproduction makes today's production facilities to shift to new concepts [12]. Such concepts are often investigated with the term "smart factory" within the era of Industry 4.0. Smart factory implies greater flexibility in production, the real-time data transfer, and the design of personalized products [12]. Thus, flexible production demands a high degree of adaptability and agility. This is supported by the integration of "assistants" such as the Internet of Things (IoT), which increases the connectivity and information transfer as well as cloud manufacturing (CM), which enables new infrastructures and service opportunities [13]. This leads to a rethinking of the product and process development, and production [3, 14].

One of the relevant processes is assembly. Assembly systems link stations, buffers, transport systems and other components. These are divided into specific subsystems or modules. The concept of an assembly system is geared to requirements [12]. The dedicated assembly system (DAS) is almost replaced by new concepts for reconfigurable and flexible assembly systems. The reconfigurable assembly system (RAS) is designed for quick adaptations of structure, hardware, and software as well as the required product capacity. The assembly system is efficient, if required changes can be forecast in advance [12]. A flexible assembly system (FAS) describes the reinforced variant of RAS with lower production volume. Main focus of FAS is on a wider range of product variants [12]. To achieve this flexibility, redundancies arise in the system, which then result in cost disadvantages [12, 15].

The extreme variant of FAS, which focuses completely on product variants, is described as the manual flexible assembly system (MFAS). In this system, workers handle different variants of products and parts. MFAS strongly influences the worker's performance, productivity as well as comfort. As shown in Fig. 1, the variance of product and the ability of mass production are significantly in contradiction. Despite its high costs, larger companies apply increasingly the principles of FAS and MFAS in their production facilities and ensure their competitiveness in terms of customer-oriented requirements.

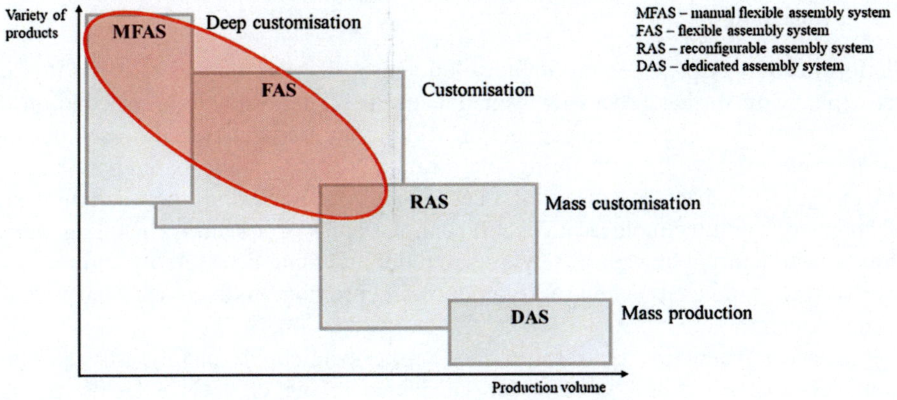

Fig. 1. Research scope of this paper—adapted from [15]

Current trends and regulations force new value creation patterns such as closed loops. At the same time, product variety increase due to customer requirements. SME can hardly effort to meet both challenges simultaneously [12]. The Urban CIRCLAS concept provides a platform for freelancers and SME. It enables a low-cost development and testing environment for assembly systems. Thus, the focus is MFAS and FAS with high flexibility and productivity. Disadvantages of these systems have to be eliminated.

2.4 Platform Economy

A platform economy is characterized by the exchange of service providers and service users. Well-known companies in commerce or travel exemplify the advantages of platform economy during the pandemic situation [16, 17]. Therefore, platform economy follows specific principles for stakeholders' interactions. For instance, stakeholders are connected and interact during 1) value creation in terms manufacturing, 2) use and consumption, and 3) reverse activities and compensation, e.g. a return service from stakeholder A to stakeholder B. This principle enables different kinds of services and products [18].

In production technology, these characteristics of the platform economy are not established yet [13]. A platform economy for an assembly system involves greater challenges. It requires a certain know-how and the equipment with the adequate utilities [19]. Efficient control of equipment and processes must be ensured. Additionally, regardless of the sector, it must be well considered who is targeted and how efficient exchanges between service providers and service users are ensured.

As it was already evident from the previous chapters, the Urban CIRCLAS concept as a platform economy is not only used for the exchange of different stakeholders. It provides a platform for testing and combining approaches with regard to new assembly systems, an urban production as well as a CE paradigm.

2.5 Research Focus

Platform economy provides profitable solutions for distribution and sale, but it is not yet explored in further processes in manufacturing. The available knowledge on the circular platform economy in production technology particularly addresses the topics of cradle-to-grave manufacturing, platform architecture and IT-based parts in manufacturing. To realize the closed-loop concept in manufacturing, there is still a lack of knowledge and technological capability in the areas of assembly, handling, MRO, cloud manufacturing as well as reverse assembly including disassembly and reassembly. Narrowing this gap is helpful to produce and promote products with multiple life cycles in the market.

An urban production is useful to shorten the local supply chain costs and emissions, but expensive and small space capacities in big cities is a challenge. The principle of platform economy provides an opportunity to meet this challenge. This principle aims to combine new assembly systems such as the FAS and MFAS and the CE in urban areas. The combination raises the interest of companies to perform CE applications for different stakeholders efficiently. This motivates the proposal of a circular platform economy for manufacturing in this paper.

3 Scope of the New "CIRCLAS" Framework

The emergence of circular economy and urbanization require a significant change in production systems including handling and assembly. Platform economy provides an approach to fulfil these requirements. This chapter outlines research fields, requirements and possible value streams. This chapter describes the Urban CIRCLAS framework and introduces both components the Urban CIRCLAS concept as well as the Urban CIRCLAS Lab within its urban space.

3.1 Requirements and Value Stream

The envisioned concept of Urban CIRCLAS relies on granular, regionally configurable production resources that achieve economic scaling effects through globalizing platform economies. The Urban CIRCLAS framework describes the realization of the Urban CIRCLAS concept. Based on the research gap, requirements for the Urban CIRCLAS framework are first established. Furthermore, the expected value streams in the Lab are specified to highlight possible fields of research.

The concept provides new solutions for industrial knowledge and applied research in the field of platform-based production. Main topics are assigned to assembly, life cycle services (repair, overhaul, product refurbishment), disassembly and reassembly. A new perspective is the sustainability-oriented character of platform-supported production. It highlights how users such as freelancers or SME combine and upgrade their know-how and capabilities to realize innovative product and process designs, which meet established industry standards.

The resilience and sustainability of production demand a rethinking of production technology in circular economy principles [10]. Material use must be developed from a cradle-to-grave to closed loops. Solution approaches for closing the loop are 1) localization of production steps and value cycles, 2) sustainability-compatible rationalization and scaling, 3) intensive maintenance, repair and upgrading, and 4) fair, network-like participation on a broad social basis.

To realize these approaches, various value streams must be ensured in the Urban CIRCLAS concept. The platform must be attractive for various stakeholders and work profitably. Figure 2 illustrates the value streams to a set of sample stakeholders ensuring an exchange for their individual interests. To serve these interests, the Urban CIRCLAS framework must be able to provide expertise as well as resources. Resources include the operating resources, IT infrastructure as well as the relevant personnel with the appropriate know-how.

The Urban CIRCLAS framework thus offers several interfaces to expand ideas, concepts and processes. The various interfaces also provide an opportunity to analyse the market from a customer, developer, investor and producer perspective. The platform enables interactive exchange about market as well as feasible developments. It also continuously structures knowledge and enhances competences as a business consultant in order to guide stakeholders. Based on the requirements, interfaces and corresponding value streams, the next chapter will explain the concept of CIRCLAS framework.

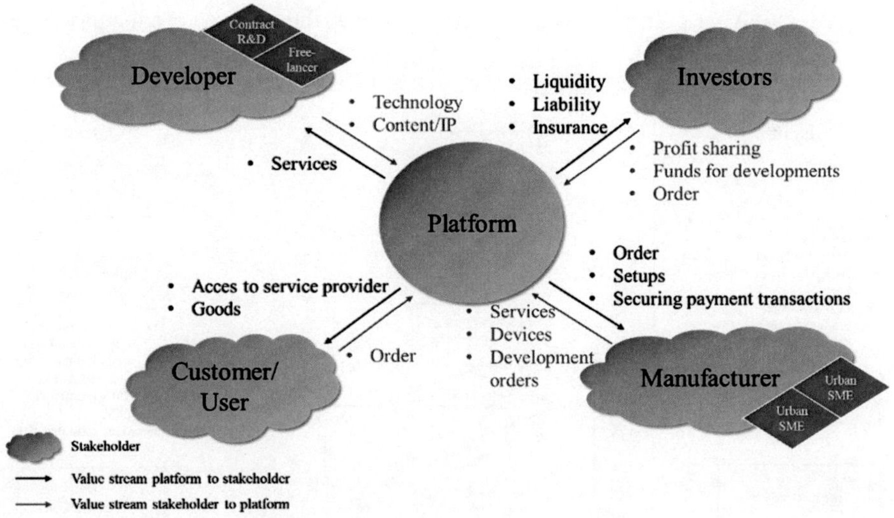

Fig. 2. Determination of stakeholder and value streams

3.2 Urban CIRCLAS Framework

The Urban CIRCLAS framework represents a production system according to the principle of platform economy. The goal is granular production resources under consideration of sustainability effects of a CE. In order to achieve this goal, it applies the previously described principles of platform economy in selected fields of production such as assembly, life cycle services and disassembly.

The key aspect in this context is that a production process is implemented at n locations and can be executed 1 time each. Despite the implementation effort at the participating sites, the effects of large quantities are aimed for. According to this principle, an ecosystem of centralized and granular or decentralized production resources are linked to relevant stakeholders via a platform economy. This ecosystem includes systems such as MFAS and FAS. Figure 3 provides an overview of the general process of the Urban CIRCLAS concept. A wide variety of requests are accepted in the Lab. The orders can be prototypes, repair measures or also testing of new resources and processes.

The left side of Fig. 3 shows input parameters including available competencies and resources such as equipment or computational tools for multiple stakeholder requests. Based on this the CIRCLAS Lab provides services on different levels with regard to innovations such as new disassembly methods, processes or newly designed products. Design for disassembly or design to cost are sample methods.

The middle part of Fig. 3 illustrates process steps, from an order to the final product. Each order is divided into suborders and the corresponding process steps are defined. For each process step, it must be checked whether the Urban CIRCLAS framework can realize the process step. Thereby not only the existing equipment is considered, but also the possibility of development and expansion of equipment and processes in the lab. If no possibilities are given, the urban location is an advantage. In the immediate area, possible production facilities for the respective production step can be considered. After going through these steps n times, the required product is created.

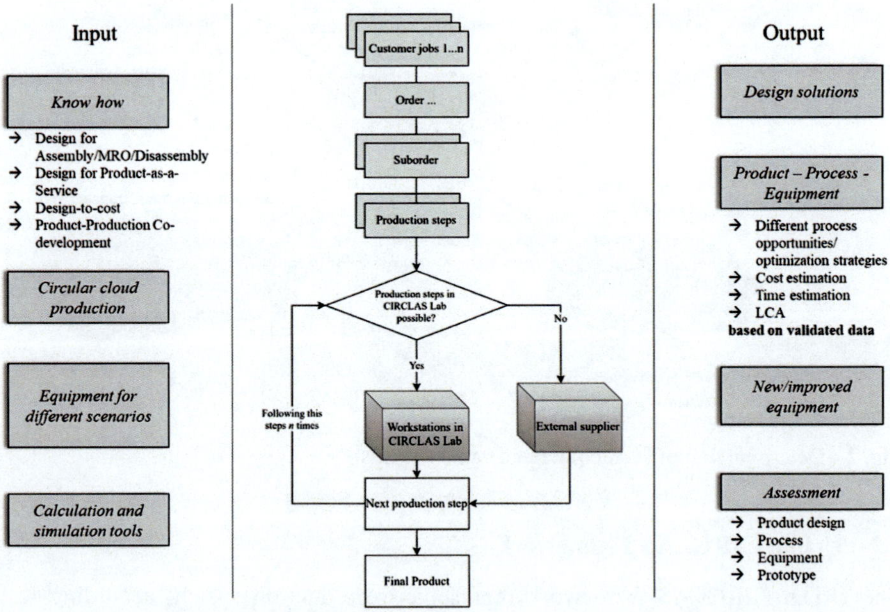

Fig. 3. Process and added value of urban CIRCLAS Lab

In addition to the process and finalization of the order, further output parameters become relevant, which are shown in the right side of Fig. 3. Different variables of the product, process and equipment can be analysed. If necessary, new solution approaches are developed. The Urban CIRCLAS framework enables the prototypical development of new equipment, which can be tested and validated directly on site in several scenarios. Due to the fact that the database is filled with real process data over multiple orders, the outcome regarding cost estimation or life-cycle assessment (LCA) are reliable. Thus, on the one hand, the platform is used for the interactive exchange of different stakeholders and, on the other hand, an innovation platform is created by identifying gaps and missing tools or resources in the different processes. The following case study illustrates the realization of change requests via the Urban CIRCLAS framework.

4 E-Scooter Case Study

A case study including multiple application scenarios demonstrate and verify the feasibility of the proposed Urban CIRCLAS framework. The E-scooter is used as a product for verification. E-scooters have become an essential vehicle for many people in big cities. Inhabitants or visitors of cities use e-scooters as a private property or through a sharing economy company, which makes the e-scooter a viable case study [20]. A repair scenario for e-scooter is used to demonstrate the advantage of the Urban CIRCLAS framework with a simple process. With regard to the strong networking to different stakeholders for e-scooters, value streams of the Urban CIRCLAS concept are specified in this chapter.

Figure 4 illustrates the value streams of the Urban CIRCLAS framework as a platform for stakeholders of e-scooters. The stakeholders are represented by 1) a manufacturer, 2) a sharing economy/private person as a customer, and 3) a developer for design and service apps of e-scooters. It is important to guarantee that data is exchanged between the platform and the stakeholders under the condition of confidentiality. This information will be processed accordingly in the database. Thus, an interactive exchange between stakeholders and the Urban CIRCLAS framework is ensured.

The central location of the CIRCLAS Lab provides a direct contact to the customer as well as to the product. This is advantageous from both customers' and suppliers' point of view. For example, all e-scooters within a 5 km radius to the CIRCLAS Lab can be automatically assigned to the CIRCLAS Lab as a service provider for maintenance. Networking with other labs in the immediate area for realizing the requests reduces transport distances and thus offers new logistics concepts. This way, a sharing economy does not have to collect these e-scooters, transport them to the corresponding repair sites and distribute them again later. Thus, logistics and transport costs for customer like sharing economies are reduced. The principle of urban production applied in a platform economy thus delivers a positive impact in terms of effort, sustainability and costs.

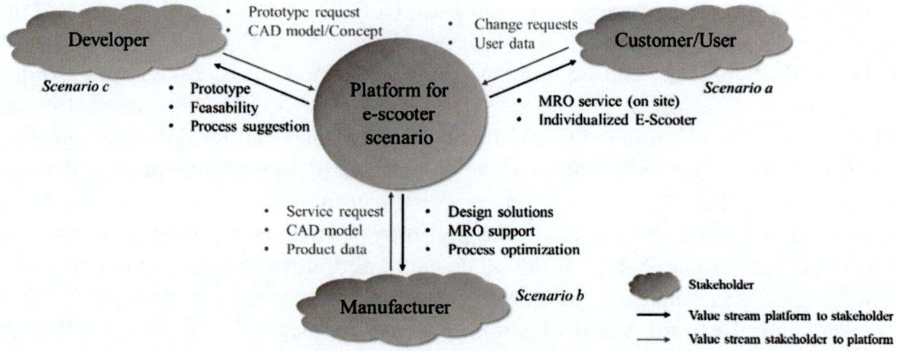

Fig. 4. User scenarios in the Urban CIRCLAS framework as a platform

The interactive exchange is shown through the value streams, which depend on the stakeholder request. In order to examine more specifically the procedure and the possibilities in the developed concept, the shown stakeholder requests will be carried out. For this purpose, Fig. 4 illustrates the value stream between a customer and CIRCLAS, for instance, in the case of repair.

(a) *E-scooter scenario between a User and the Urban CIRCLAS framework*

The change request is, as mentioned, the replacement of the front mudguard on the first wheel of a conventional e-scooter, which is illustrated in Fig. 5. Due to the fact, that the stakeholder is a user, less product data is available. The user is only able to provide his user data. Thus, based on the existing know-how, the required functionalities can be defined and a replacement part for the e-scooter can be developed and transferred in a CAD model. This CAD model in turn is used for 3D printing. The new component is used to replace the damaged part. For this, two options to replace this front mudguard are possible the 1) replacement of the mudguard in the CIRCLAS Lab or a 2) replacement of the mudguard directly at the damaged e-scooter (on-site or remote).

Depending on the change request, a repair in the CIRCLAS Lab can be carried out manually or partially automated. This is enabled by using the concepts of FAS and MFAS. The decision depends on the workload caused by other requests. The availability of a service employee as well as the repair equipment is essential. In this customer request, it is assumed that an on-site repair is needed.

In aviation and other sectors, the development of a remote guidance for maintenance has become important [21, 22]. Especially in times of the pandemic, it became relevant to support repair remotely. Cloud manufacturing (CM) supported approaches are being explored, which ensures efficient data exchange. Using CM-supported approaches, for example, enables to remotely identify the problem and determine the appropriate repair action. Guidance via augmented reality (AR) supports to execution of the determined repair actions [23]. The Urban CIRCLAS framework tests CM-supported approaches and remote maintenance service for any sector. Thus, the Urban CIRCLAS framework provides new equipment for different services and it tests how much information in the remote maintenance service is relevant. Therefore, it extends the idea of a remote maintenance service.

Regarding to the e-scooter scenario, a sharing economy benefits from CM and performing condition monitoring. All e-scooters in the immediate vicinity can be serviced, when a specified period of time has elapsed since the last maintenance or certain performance parameters indicate possible faults. The possibility of monitoring as well as appropriate response are offered. Thus, the Urban CIRCLAS framework realizes repairs at an early stage and in case of frequent errors, appropriate design solutions are generated.

If the presence of employees and corresponding equipment becomes necessary, further aspects have to be considered. It has to be ensured that the service can be carried out with low effort. An on-site repair imposes new requirements on the equipment. Thereby the weight of the equipment is relevant as well as the number of equipment that has to be transported. The savings in the transfer of employees and equipment offer cost reductions, for instance, in the maintenance of offshore wind turbines (OWT). The case study shows that the equipment must be appropriately adapted to different conditions. Through CIRCLAS, equipment can be developed, tested in real scenarios, and evaluated accordingly [24].

(b)*E-scooter scenario between a manufacturer and the Urban CIRCLAS framework*

The example e-scooter, which has a damaged front mudguard, is further used. As illustrated in Fig. 4, a manufacturer provides product data and CAD models along with its service request. The manufacturer is interested in effective design solutions on the product.

Due to the close proximity of the CIRCLAS Lab to the product, a quick identification of the damage and the frequency of the damage is enabled. The front mudguard as a part, which is connected to the control arm, has to be replaced. This means that a replacement of the entire mudguard is necessary. However, since the front part of the mudguard has correspondingly more often foreign damage due to its position, the rear part of the mudguard is usually undamaged. A design for disassembly principle can ensure faster replacement. Manufacturing the front mudguard in two parts, is a possible design solution as shown in Fig. 5. These parts can then be connected with a screw connection. The front part, which is more often damaged, can be mounted and dismounted directly to the rear mudguard part with two screws. This ensures a replacement of only the damaged part and resulting in material savings. These optimizations can already be observed in newer e-scooters, resulting in significant advantages.

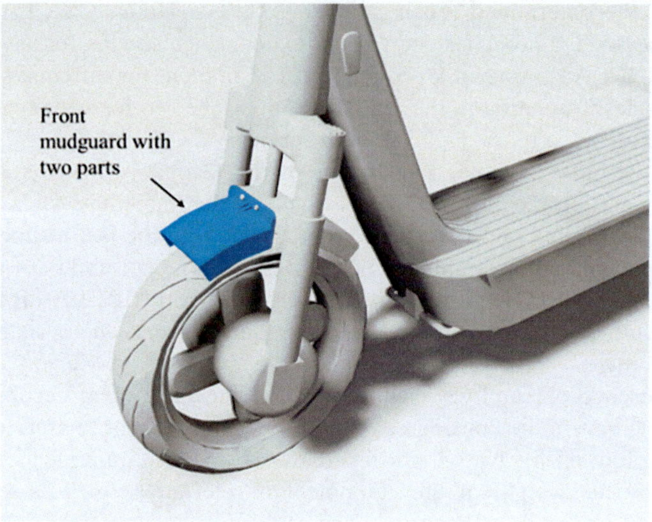

Fig. 5. Front mudguard with screw connection

This example shows that the design for disassembly approach can reduce costs through material savings on the product as well as time savings in the process. In addition, there is the sustainability aspect, which is ensured by reduced waste.

(c) *E-scooter scenario between a developer and the Urban CIRCLAS framework*

A developer can be a freelancer or an SME and does not have many resources or corresponding competencies for the manufacturing process. The missing resources as well as competences are offered in the Urban CIRCLAS framework. This results in the value streams, which are shown in Fig. 4. The developer has a prototype request and a concept or ideally a CAD model. The required process steps are derived from the CAD model. The necessary operating resources are then assigned to the respective process steps. Due to the continuous collection of data regarding the own process times, an estimation of the time for the process can be made. A first cost estimation is made based on machine hour rates. Developed tools for the cost estimation as well as the LCA are used. The developer gets an overview of his product idea from different perspectives, whereby the research fields in the Urban CIRCLAS framework expands.

Based on the e-scooter scenario, the developer asks for a prototype to demonstrate lower crash rates. For example, the developer expects a lower crash rate with an e-scooter that has two wheels at the back and one at the front. This should reduce the need for repairing the front mudguards and also minimize the risk of injury to the user. At this point, an initial evaluation can be made in the early stages of product development using virtual engineering (VE) [25]. The Urban CIRCLAS framework offers an implementation of this concept throughout the entire life cycle. It is thus guaranteed that by means of real data and an implementation of VE, conclusions can be drawn about the product and process.

Summary
The application of three scenarios on the proposed Urban CIRCLAS framework demonstrates the value streams between different stakeholders. The positive impact of the Urban CIRCLAS framework caused by the given opportunities such as competencies in design methods or equipment and the possibility to develop equipment, are elaborated in the scenarios. In summary, the Urban CIRCLAS framework provides in these scenarios 1) (remote) maintenance service despite no product data, 2) design solutions, which result in reduction of material waste and costs, and 3) prototyping using AR and VE, which enables conclusions and forecasts. However, it is evident from the three scenarios that the Urban CIRCLAS framework has to react quickly. This fact promises a quick identification of weaknesses in the equipment as well as in the system set up. For example, a constant retooling of gripper systems does not offer any direct added value and is associated with time and costs. The gripper system in the e-scooter scenario must be constantly changed if new requests regarding the handlebars, battery or wheel arise in addition to a mudguard change. Therefore, this is a field of research that ensures a minimization of changeovers. In this sense, the sustainability aspect is also ensured, since less material is indirectly consumed due to a lower number of equipment accessories. In addition, adaptability to new product applications and variants can be expanded. Thus, weaknesses such as gripper system are identified based on the implementation of various services.

5 Conclusion and Outlook

This paper investigates the development of a platform economy principle for manufacturing that meets today's requirements and trends such as mass customization and urban production. The proposed concept addresses the challenge of combining different disciplines in the field of assembly, handling, disassembly and reassembly. This paper promotes the framework "Urban Circular Cloud Assembly and Services" (Urban CIRCLAS), which consists of a new concept for platform economy applications and of a lab for creation and demonstration of granular and order-specific ad-hoc new combination of operating resources, processes and workers.

The goal of the Urban CIRCLAS framework is creation of new perspectives through networking of different stakeholders. In this context, a database for cloud manufacturing as well as generic tools are developed in a platform that offers forecasts and justifies effort and costs for different scenarios. In sample scenarios, maintenance intervals can be predicted, process costs estimated or product developments evaluated.

Based on the scenarios, various benefits for stakeholders are identified such as providing competences and the urban location. Further research fields emerge to be explored within the Urban CIRCLAS Lab:

- Cloud manufacturing and condition monitoring to reduce repair time,
- Equipment development for on-site maintenance,
- Design solutions using principles such as design for disassembly,
- Reducing material waste and time waste in repair and overhaul processes,

- Cost reduction in MRO and prototyping,
- Virtual engineering for sustainable manufacturing,
- LCA for the standardization of products and processes, and
- Prototyping for micro mobility solutions.

Research and industrial activities in these fields focus on isolated developments and applications, but have not been widely combined. The Urban CIRCLAS framework combines multiple aspects, especially advantages of flexible assembly systems, mass customization and sustainability consideration in a single multifaceted platform. Stakeholders, services and competences can interact in various ways within the platform, which connects the interactions and leads fast and meaningful conclusions.

The CIRCLAS Lab provides a physical environment for research and development of pioneer solutions based on the platform characteristics of the framework. The framework offers continuously updated services. Urbanization as well as networking are essential components of this framework, which support fundamental research projects and innovative applications. The Urban CIRCLAS framework creates a set of technical requirements and solutions for ensuring quality and trust in products and services via the platform virtually as well as the Lab physically.

With regard to circular economy, the Urban CIRCLAS framework enables both forward and reverse processes. As a main part of end-of-life, separation processes are investigated in order to design the recycling process in a low-cost and value-adding way. As a case study, a second-life scenario is created by separating the parts or components of a e-scooter appropriately. This case study exploits the limits of the continuous reuse of raw materials.

Future work aims to implement the Urban CIRCLAS framework in the Lab. Identified research fields will be investigated to develop new methods, tools and equipment enhancing benefits of Urban CIRCLAS framework.

References

1. Wallis, H., Loy, L.S.: What drives pro-environmental activism of young people? A survey study on the fridays for future movement. J. Environ. Psychol. **74**(101581), 1 (2021). https://doi.org/10.1016/j.jenvp.2021.101581
2. The European green deal. European Commission, Brussels (2019)
3. Battaïa, O., Otto, A., Sgarbossa, F., Pesch, E.: Future trends in management and operation of assembly systems: From customized assembly systems to cyber-physical systems. Omega **78**, 1–4 (2018). https://doi.org/10.1016/j.omega.2018.01.010
4. Reichler, A.-K., et al.: Incremental manufacturing: Model-based part design and process planning for hybrid manufacturing of multi-material parts. Procedia CIRP **79**, 107–112 (2019). https://doi.org/10.1016/j.procir.2019.02.020
5. Baranauskas, G.: Digitalization Impact on transformations of mass customization concept: Conceptual modelling of online customization frameworks. Mark. Manag. Innov. **3**, 120–132 (2020). https://doi.org/10.21272/mmi.2020.3-09

6. Burggräf, P., Dannapfel, M., Uelpenich, J., Kasalo, M.: Urban factories: Industry insights and empirical evidence within manufacturing companies in German-speaking countries. Procedia Manufac. **28**, 83–89 (2019). https://doi.org/10.1016/j.promfg.2018.12.014
7. Herrmann, C., Juraschek, M., Burggräf, P., Kara, S.: Urban production: State of the art and future trends for urban factories. CIRP Ann. **69**(2), 764–787 (2020). https://doi.org/10.1016/j.cirp.2020.05.003
8. Johansson, P.E.C., Malmsköld, L., Fast-Berglund, Å., Moestam, L.: Challenges of handling assembly information in global manufacturing companies. J. Manuf. Technol. Manag. **31**(5), 955–976 (2020). https://doi.org/10.1108/JMTM-05-2018-0137
9. Suzanne, E., Absi, N., Borodin, V.: Towards circular economy in production planning: Challenges and opportunities. Eur. J. Oper. Res. **287**(1), 168–190 (2020). https://doi.org/10.1016/j.ejor.2020.04.043
10. Dey, P.K., Malesios, C., De, D., Budhwar, P., Chowdhury, S., Cheffi, W.: Circular economy to enhance sustainability of small and medium-sized enterprises. Bus. Strateg. Environ. **29**(6), 2145–2169 (2020). https://doi.org/10.1002/bse.2492
11. Matt, D.T., Orzes, G., Rauch, E., Dallasega, P.: Urban production—A socially sustainable factory concept to overcome shortcomings of qualified workers in smart SMEs. Comput. Ind. Eng. **139**, 105384 (2020). https://doi.org/10.1016/j.cie.2018.08.035
12. ElMaraghy, H., ElMaraghy, W.: Smart adaptable assembly systems. Procedia CIRP **44**, 4–13 (2016). https://doi.org/10.1016/j.procir.2016.04.107
13. Gerrikagoitia, J.K., Unamuno, G., Urkia, E., Serna, A.: Digital manufacturing platforms in the industry 4.0 from private and public perspectives. Appl. Sci. **9**(14), 2934 (2019). https://doi.org/10.3390/app9142934
14. Cohen, Y., Faccio, M., Pilati, F., Yao, X.: Design and management of digital manufacturing and assembly systems in the Industry 4.0 era. Int. J. Adv. Manufac. Technol. **105**(9), 3565–3577 (2019). https://doi.org/10.1007/s00170-019-04595-0
15. Cohen, Y., Faccio, M., Galizia, F.G., Mora, C., Pilati, F.: Assembly system configuration through Industry 4.0 principles: The expected change in the actual paradigms. IFAC-PapersOnLine **50**(1), 14958–14963 (2017). https://doi.org/10.1016/j.ifacol.2017.08.2550
16. Kenney, M., Zysman, J.: The platform economy: Restructuring the space of capitalist accumulation. Camb. J. Reg. Econ. Soc. **13**(1), 55–76 (2020). https://doi.org/10.1093/cjres/rsaa001
17. Dermott, M.: Robots and AR: Towards a platform economy for construction. ITcon **24**, 527–539 (2019). https://doi.org/10.36680/j.itcon.2019.029
18. Drewel, M., Özcan, L., Gausemeier, J., Dumitrescu, R.: Platform patterns—using proven principles to develop digital platforms. J. Knowl. Econ. **12**(2), 519–543 (2021). https://doi.org/10.1007/s13132-021-00772-3
19. Leal Filho, W., Azul, A.M., Brandli, L., Özuyar, P.G., Wall, T. (Eds.): Affordable and clean energy. Springer, Cham (2019)
20. Nikiforiadis, A., Paschalidis, E., Stamatiadis, N., Raptopoulou, A., Kostareli, A., Basbas, S.: Analysis of attitudes and engagement of shared e-scooter users. Transp. Res. Part D Transp. Environ. **94**(102790), 1–14 (2021). https://doi.org/10.1016/j.trd.2021.102790
21. Utzig, S., Kaps, R., Azeem, S.M., Gerndt, A.: Augmented reality for remote collaboration in aircraft maintenance tasks. IEEE Aerospace Conference, Piscataway (2019). http://ieeexplore.ieee.org/servlet/opac?punumber=8727865
22. Vorraber, W., Gasser, J., Webb, H., Neubacher, D., Url, P.: Assessing augmented reality in production: Remote-assisted maintenance with HoloLens. Procedia CIRP **88**, 139–144 (2020). https://doi.org/10.1016/j.procir.2020.05.025

23. Mourtzis, D., Siatras, V., Angelopoulos, J.: Real-time remote maintenance support based on augmented reality (AR). Appl. Sci. **10**(5), 1855 (2020). https://doi.org/10.3390/app10051855
24. Ren, Z., Verma, A.S., Li, Y., Teuwen, J.J., Jiang, Z.: Offshore wind turbine operations and maintenance: A state-of-the-art review. Renew. Sustain. Energy Rev. **144**(110886), 1–22 (2021). https://doi.org/10.1016/j.rser.2021.110886
25. Assad, F., Konstantinov, S., Rushforth, E.J., Vera, D.A., Harrison, R.: Virtual engineering in the support of sustainable assembly systems. Procedia CIRP **97**, 367–372 (2021). https://doi.org/10.1016/j.procir.2020.05.252

Design and Simulation

AI-Based Performance Prediction and Its Application on the Design and Simulation of Cooling Plates for Battery Electric Vehicles

Niklas Klinke[1(✉)], Stefan Buchkremer[1], Lutz-Eike Elend[1], Maksym Kalaidov[2], and Thomas Tschammer[2]

[1] Mubea New Body Products, Mubea-Platz 1, 57439 Attendorn, Germany
{niklas.klinke,Stefan.Buchkremer, Lutz-Eike.Elend}@mubea.com

[2] Neural Concept, EPFL Innovation Park Building C, 1015 Lausanne, Switzerland
maksym.kalaidov@neuralconcept.com

Abstract. With the increasing focus on the electrification of personal mobility, shortened development cycles with high cost pressure have to be managed quicker than ever in the Automotive Industry at minimal development cost. As a global automotive supplier, Mubea develops novel manufacturing technologies for new products in the electric powertrain. One example in the battery case of battery electric vehicles is the thermal management system of the traction batteries—a key factor in the battery case. As a new production technology for this product, the Mubea Rollbonding Process of aluminum offers several advantages, such as a high design freedom of the channel structure. For the product development—even though Mubea consequently makes use of automated simulation workflows—the turn-around time of a single CFD performance evaluation is still high. Therefore, our goal is to use all the historic simulation results from past projects to build a predictive model that allows the prediction of simulation results in real-time. However, given the high freedom in the design space allowed by the Rollbonding process, standard Machine Learning approaches, based on parameters, are not suitable. Hence, there is the need to directly process 3D geometries as such. Using historical engineering data, the unique deep learning approach of Neural Concept is able to predict unseen designs in seconds rather than hours, directly from the raw CAD file. This innovative approach allows Mubea to iterate faster and shorten the response time on customer enquiries. In conjunction with other design disciplines and manufacturing data, we look forward to have not only an AI-based design evaluation but also a tolerance-aware design optimization. In this paper we present an innovative strategy to utilize historic simulation results, and the corresponding 3D geometries, to predict the performance of new designs instantaneously. After explaining the underlying approach, first results are discussed. It can be shown that with as little as 100 training samples, this approach is able to deliver predictions with sufficient accuracy and over 90%

K. Dröder and T. Vietor (Eds.): *Future Automotive Production Conference 2022*, Zukunftstechnologien für den multifunktionalen Leichtbau, pp. 207–218, 2023.
https://doi.org/10.1007/978-3-658-39928-3_15

of lead-time reduction. Finally, we explain how Neural Concept and Mubea are collaborating to embed this approach in the Mubea design and simulation environment.

Keywords: Machine learning · CFD-simulation · Design and simulation · Rollbonding · Cooling plates · Battery electric vehicles

1 Introduction

Rollbonding (RB) is a manufacturing process where two or more metallic sheets are welded together by a flat rolling process. The bonding can be performed at room temperature or after a heat-treatment—in warm condition [1]. Applications of this process range from production of bimetal sheets for electric engine components to manufacturing of heat exchangers. Since it is possible to print a separator medium on one of the sheets with a pattern the bonding can be controlled: Only the bare metal surfaces are bonded, while the painted surface remains separated [2]. The joined plate is inflated by applying pressurized air after the rollbonding, which forms out a channel geometry. Figure 1 shows the basic principle.

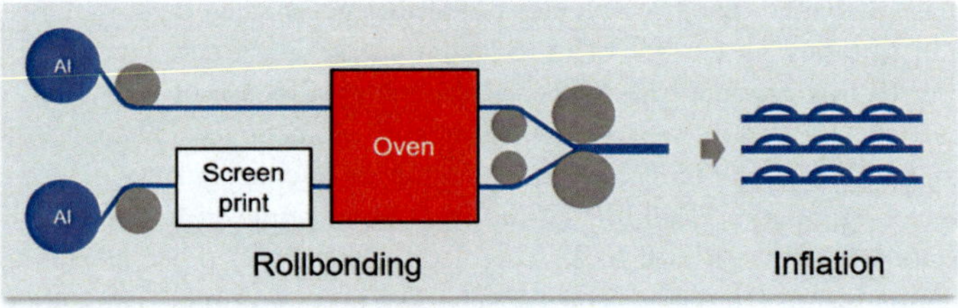

Fig. 1. Schematic illustration of the manufacturing of rollbonded plates

With increasing focus on electrification of personal mobility, the commonly known manufacturing process for freezers and refrigerators was advanced by Mubea to fulfill the requirements of automotive thermal management systems. The Mubea rollbonding process delivers thermal management systems that are characterized by the highest cooling and heating capacities as well as outstanding economy and maximum design freedom. The channel design can be varied in topology as well as in geometry. As shown in Fig. 2 the topology can be anywhere between a dimple design and classical channel design. The high design freedom makes it possible to truly optimize the cooling systems performance, but requires several dozen of design iterations.

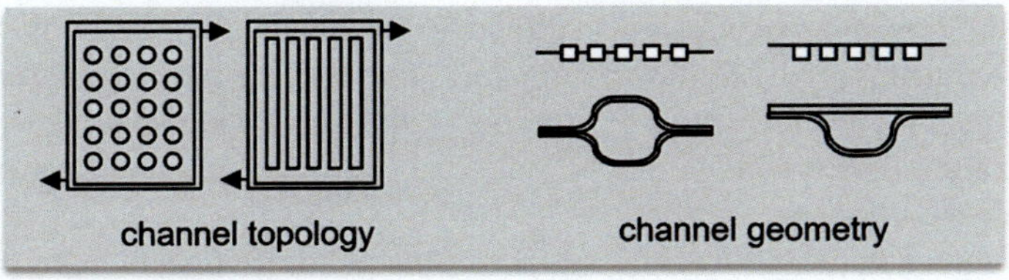

Fig. 2. Design freedom of rollbonded plates

To design the cooling channels adequately, the thermo-hydraulic behavior is simulated using computational fluid dynamic (CFD) codes. The designer evaluates key performance indices (KPIs) like the maximum temperatures, fluid velocities and pressure drop throughout the design process. Due to the shape of the cooling plate (approx. 1500 mm x 2000 mm x 3 mm), a simulation model that covers all details consists of around 30 million cells. To reach a steady state numerous iterations have to be calculated by the CFD solver on a high performance computing cluster.

Since 2015 Mubea continuously works on the application of optimization methods on the design of body components [3]. We identified the automated design evaluation as huge synergy between manual and automatic design optimization. Hence, we democratized simulation methods and enabled our designers to evaluate performance without the need of becoming a CAE expert. Even though our designers can evaluate on themselves, the turnaround time is in the range of a day. This is why a quicker design evaluation becomes necessary to enable automatic design optimization and faster manual design.

Classical design optimization is based on parameterized designs and the evaluation of scalar KPIs. Design variables are expressed as vector $\boldsymbol{x}$, while responses are denoted as vector $\boldsymbol{y}$, with $x, y \in R^n$. The general relationship between design variables and responses can be expressed as:

$$\boldsymbol{y} = f(\boldsymbol{x}) \tag{1}$$

Here f is the simulation model. For computational demanding tasks, optimization algorithms typically are not applied to the design problem directly but to a surrogate problem [4]. A so called metamodel is trained based on machine learning algorithms to predict the KPIs by interpolating the performance landscape given a low dimensional parametrization of the shape space. This metamodel is then used as a proxy for the true objective to speed-up the computation. The altered model relationship can be expressed as:

$$\widehat{\boldsymbol{y}} = \widehat{f}(\boldsymbol{x}) \tag{2}$$

Now the simulation model is replaced by the approximation $\widehat{f}$, which delivers predicted responses $\widehat{\boldsymbol{y}}$. To train a metamodel, necessary training data is generated based on a design of experiments, where parameters are varied in a controlled manner to reach a good design space coverage [4].

However, this approach is only effective for shape deformations that can be parameterized using relatively few parameters. Their performance therefore hinges on a well-designed parameterization. Furthermore, the models are specific to a particular parameterization. Historical simulation data lacking a joint parametrization cannot be leveraged easily.

Therefore, when designing rollbonded cooling plates, two challenges arise:

1. Response distributions are of interest, hence we need a prediction of the response field rather than scalar KPIs,
2. A meaningful parametrization of the design space is nearly impossible when the design freedom of the technology shall be covered.

A common approach to tackle the first challenge is the so-called reduced order modeling (ROM), which enables the prediction of field responses. Based on an proper orthogonal decomposition (POD) the field response can be expressed with a reduced set of eigenvectors and predicted based on classical metamodeling [5–7]. The downside of this approach is that the field response is smoothed and non-linearities are not well captured. Other than that, a good parametrization is still needed.

The Neural Concept approach, based on 3D Deep Learning and Geodesic Convolutional Neural Networks (GCNN), is the first AI approach which is able to deal with unstructured 3D data from CAD geometries directly, and learns how they interact with the laws of physics. It is able to predict both global quantities and fields responses from a single geometry. As neural networks are designed to model any response function, they are also able to accurately capture even highly non-linear phenomena, or discontinuities in the physical response. Moreover, as this approach is handling raw CAD data, it is free from any underlying parametrization of the geometry. This means these predictive models can learn from many different topologies from various sources, hence handling a much wider design space. This is especially suited for use-cases with a large freedom in the design space, as enabled by the Mubea Rollbonding process.

In this work, we applied the Neural Concept Shape approach to a set of historical CFD simulations of cooling plates.

2 AI Based Preformance Prediction with Neural Concept Shape

The Neural Concept Shape approach is based on Geodesic Convolutional Neural Networks (GCNN). These neural networks models are able to deal with a 3D representation of the geometry and extract the relevant features from it. This is the result of years of research from the Computer Vision Laboratory of EPFL, later further developed and implemented by Neural Concept [8].

It uses multi-scale geometric neural networks, through a combination of surface such as geodesic and euclidean network architectures. The first part of the model pre-processes the input and constructs a set of features by means of the previously introduced geodesic convolution operations. These features are used to predict the global scalars via average pooling within two dense layers. The second branch of the network generates fields relying on an additional set of geodesic convolutions and point-wise operations. The global architecture is shown with Fig. 3

This new network takes advantage of a GPU efficient implementation of geodesic convolutions, removing the need to use a Cube-Mesh mapping or any prior remeshing.

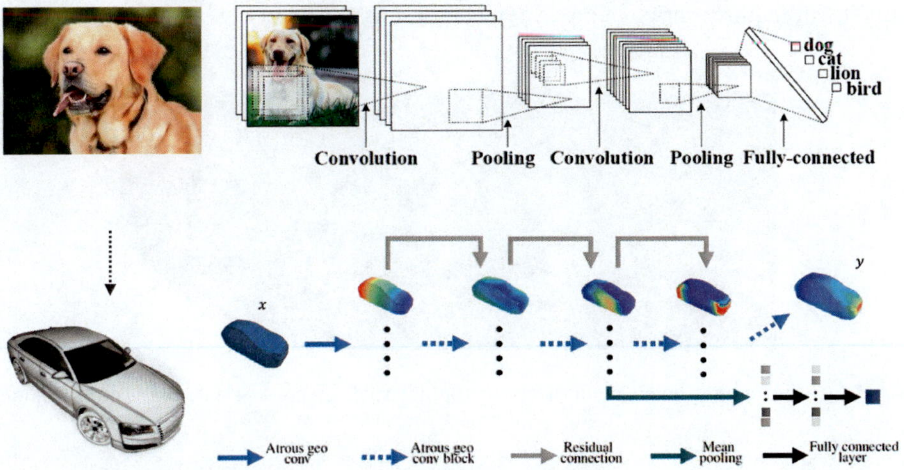

Fig. 3. Neural Concept Shape network architecture

3 Application to Cooling Plates

In the development process of cooling plates, the task of the designer is to develop a channel topology that provides an uniform temperature distribution with minimum pressure drop Δp. Module areas, heat flux Q, inlet temperature T_{inlet} as well as the mass flow rate $\dot{m}$ are given quantities. Figure 4 depicts those analysis quantities.

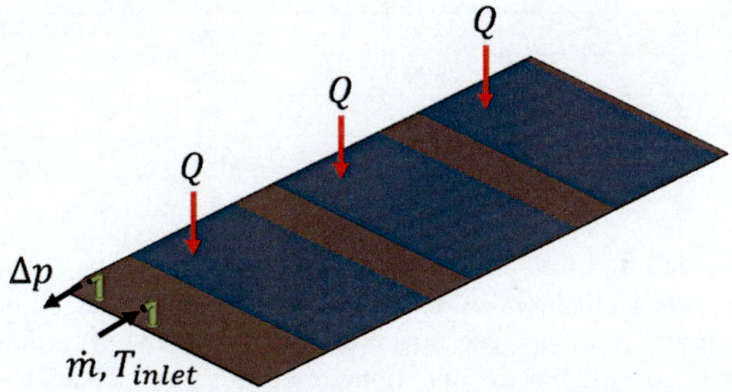

Fig. 4. Cooling plate with analysis quantities

Mubea uses Ansys Fluent to solve the underlying steady state conjugated heat transfer problem by applying a coupled solution scheme and kω-SST turbulence modeling. As important as the right specification of material properties is the appropriate meshing of the geometry. In order to correctly model the thermal behavior on the fluid wall, mesh refinement is applied. This results in a high number of elements, long meshing and solving times and huge amount of data to handle in post-processing. Even though meshing, solving and post-processing can be done without user interaction, it still requires serious time. A typical mesh is shown in Fig. 5.

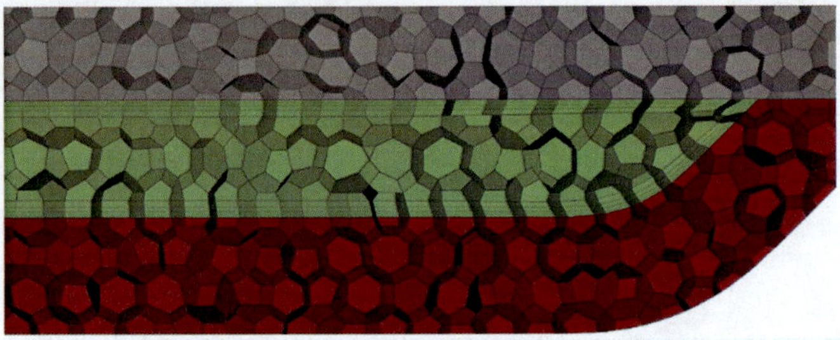

Fig. 5. Cut through a channel mesh (gray: top/flat side, green: coolant, red: bottom/channel side)

In order to speed up the design evaluations, the Neural Concept Shape algorithm was trained using historical project data. The used data originates from five different vehicle programs, whose project and design space characteristics are quite different, as shown in Table 1. In addition to the design space, also the input parameters vary between projects: the inlet mass flow $\dot{m}$ ranges from 6–16 l/min, the inlet temperature T_{inlet} is defined within a range of 288.15–298.15 K, while the module heat flux Q varies within 1.7–6.1 kW/m^2.

Table 1. Characteristics of the different vehicle programs

Vehicle program	Number of modules	Average plate length [mm]	Average plate width [mm]	Number of samples
A	9	2000	1000	25
B	5	2300	650	33
C	4	1150	500	27
D	3	800	550	33

To train a model, first the inputs $\boldsymbol{x}$ and the output quantities $\boldsymbol{y}$ have to be defined. The new aspect here is, that next to the obvious inputs, heat flux Q, inlet temperature T_{inlet} and mass flow rate $\dot{m}$, the geometry is used as it is, without further simplification. Anyhow, the geometry file has to contain the relevant areas such as the inlet and outlet surfaces, the module areas as well as the solid and fluid regions.

The aim of the model is to predict all relevant responses the designer typically analyzes such as: temperature field on the top of the plate $\boldsymbol{T}_{top}$, velocity field in the fluid midplate $\boldsymbol{v}_{fluid}$, temperature field in the fluid midplane $\boldsymbol{T}_{fluid}$, pressure field in the fluid midplate $\boldsymbol{p}_{fluid}$, pressure drop Δp, maximum top plate temperature $T_{top,max}$, maximum fluid velocity $v_{fluid,max}$ and max fluid temperature $T_{fluid,max}$.

3.1 Evaluation of Prediction Accuracy

Based on the mentioned inputs and outputs a model is trained using 108 training samples from the vehicle programs A–D. The training took about 30 h on a single NVIDIA Tesla K80 GPU. Afterwards the model was tested with 10 samples that were hold out.

To evaluate the prediction accuracy of the model, the relative mean absolute error e_{RMAE} is calculated across all n_d samples, by normalizing the mean absolute error e_{MAE} with the value range of the sample:

$$e_{MAE} = \frac{1}{n_d}\sum_{i=1}^{n_d}\left|y_i - \widehat{y}_i\right| \tag{3}$$

$$e_{RMAE} = \frac{e_{MAE}}{y_{max} - y_{min}} \tag{4}$$

Table 2 shows the RMAE for the different field responses, averaged across the samples of the programs. The error is within an appropriate range below 10% with 6.78% as maximum and 4.15% as mean.

Table 2. Field prediction accuracy of the test set averaged across different programs

Program	$\boldsymbol{T}_{top}$[%]	$\boldsymbol{v}_{fluid}$[%]	$\boldsymbol{T}_{fluid}$[%]	$\boldsymbol{p}_{fluid}$[%]
A	2.47	2.47	5.81	6.00
B	3.21	2.08	5.89	6.78
C	4.42	4.33	5.47	4.43
D	1.66	3.78	3.26	4.28

Table 3 shows the prediction errors on the scalar KPIs as absolute and relative values. For most programs, the error is far below 1%, while single responses show errors of up to 9.1%. The mean error is 2.01%.

Table 3. Scalar prediction accuracy of the test set averaged across different programs

Program	$T_{top,max}$[K]	$v_{fluid,max}$[m/s]	$T_{fluid,max}$[K]	Δp[Pa]
A	0.0481 (0.02%)	0.0742 (2.91%)	0.0820 (0.03%)	1146.8 (4.75%)
B	0.1625 (0.05%)	0.2592 (5.23%)	0.0709 (0.02%)	978.0 (3.29%)
C	0.2285 (0.07%)	0.3614 (9.10%)	0.3614 (0.01%)	412.3 (0.84%)
D	0.0763 (0.02%)	0.1583 (5.09%)	0.0742 (0.09%)	285.3 (0.71%)

To show the qualitative difference between simulation and model prediction, two characteristic test samples are selected (B20 and C14).

Figure 6 shows a good prediction with minor differences in the values on sample B20. Here the maximum error is 0.266 K (6.65%), which is fairly low. Hotspots are well captured in position and size. Most predictions in the test set are comparable to this in quality. In Fig. 7 "the worst" prediction from the test sample C14 is shown. This prediction misses several hotspots and has a maximum error of 1.35 K (9.64%).

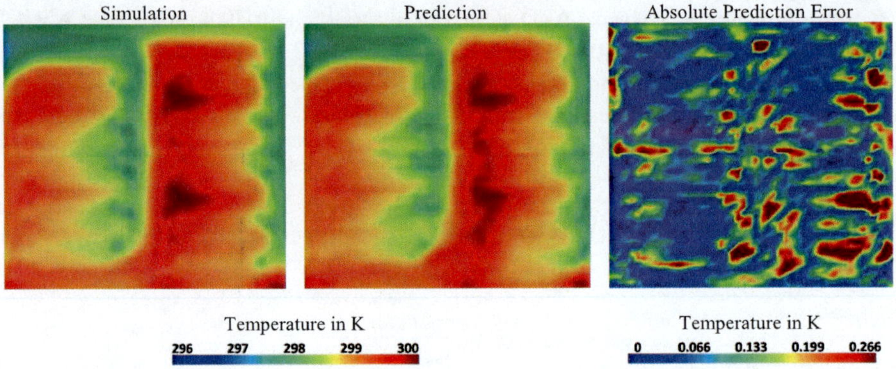

Fig. 6. Prediction of temperature field on the top of the plate $\boldsymbol{T}_{top}$ of Sample B20

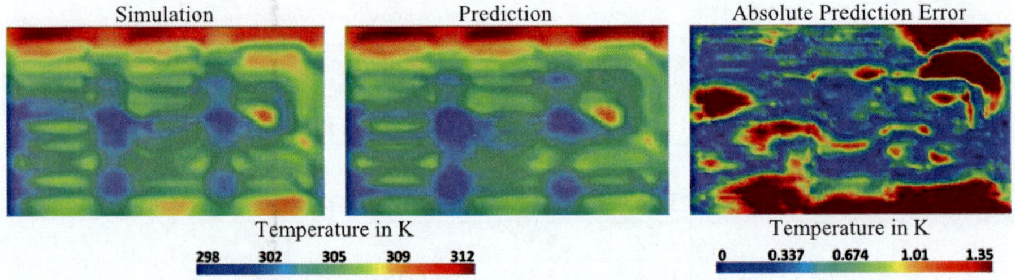

Fig. 7. Prediction of temperature field on the top of the plate $\boldsymbol{T}_{top}$ of sample C14

3.2 Evaluation of Model Uncertainty

In order to use the NCS prediction in project context rather than CFD simulations, hotspots have to be captured accurately and the prediction error should be low in general. Worst case would be when the prediction drives the design in the wrong direction. In order to judge if a prediction is reliable, the evaluation of the model uncertainty is a possible method.

To derive the model uncertainty, the model was trained in such a manner, that it was not only learning the function that maps inputs to the outputs, but it also was learning the distribution of possible outputs given a single input. Thus, during inference, the model produces an ensemble of predictions sampled from this learned distribution. While the average of this ensemble is considered to be the actual output prediction, double of its standard deviation is taken to be the 95-% confidence interval for uncertainty estimation.

Figures 8 and 9 show the prediction uncertainty on the two test characteristic test samples.

On sample B20, the uncertainty and the error do not correlate well. Areas of high uncertainty show low errors and vice versa. For sample C14 the uncertainty is high in the areas of high error and especially in regions where hotspots were missed.

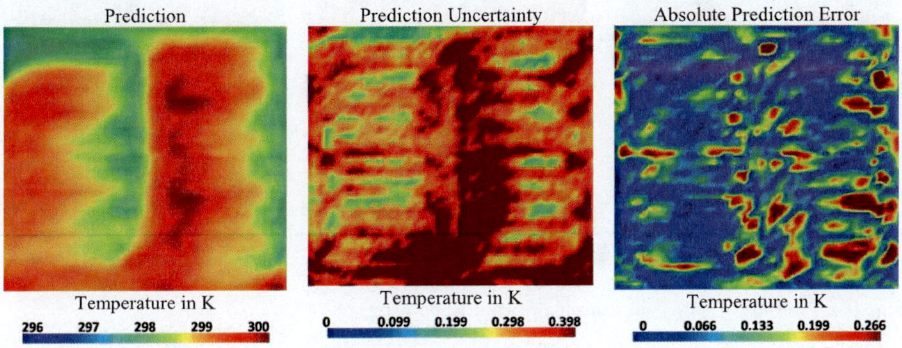

Fig. 8. Prediction uncertainty on sample B20

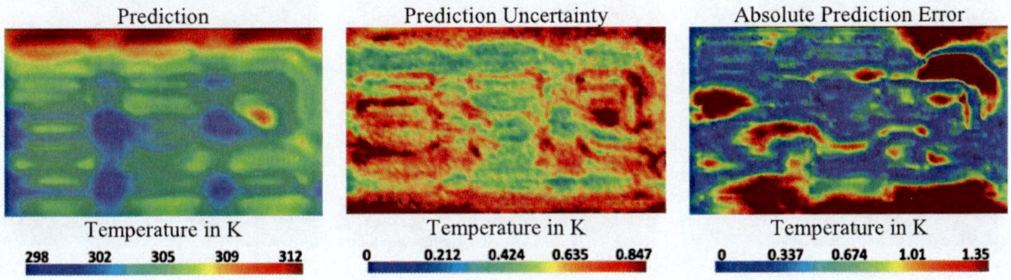

Fig. 9. Prediction uncertainty on sample C14

3.3 Evaluation of Generalization Capability

The shown results were created by using the same vehicle programs for training and validation of the model. In the industrial context, the generalization capability of the model is of high interest. Therefore, we evaluated how good the approach can predict the unseen vehicle program, by training a model with vehicles A–C and validating against D.

Because the training space of vehicle D was different to A–C, two samples from program D were used to transfer the learned model to the new boundary conditions.

After adding just two samples from the new program, the overall location of the hot-spots was predicted correctly, while the difference in the field prediction reaches 20% of the value range (Fig. 10).

After training the model with five samples from program D, the prediction got more accurate and difference in the field prediction drops below 10%, as shown in Fig. 11.

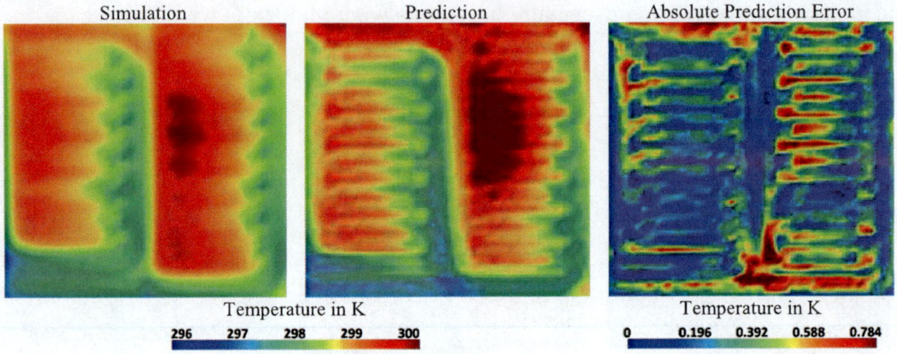

Fig. 10. Prediction with two samples from unseen program

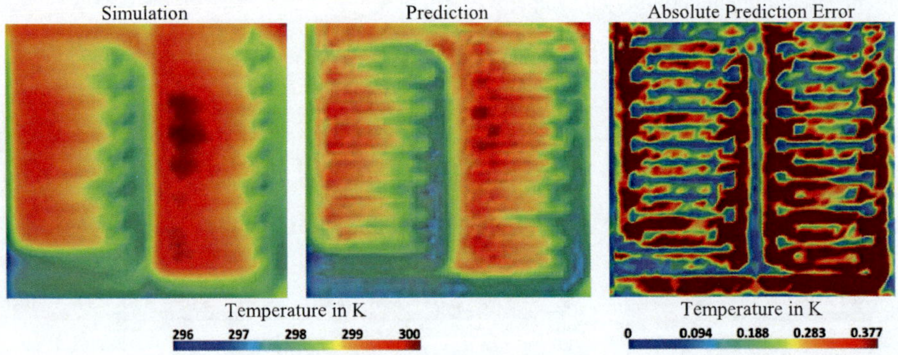

Fig. 11. Prediction with five samples from unseen program

4 Conclusion and Outlook

With the presented approach it is possible to predict full field responses based on scalar parameters in conjunction with native CAD input. This enables the re-use of historical simulation results in order to train predictive models. For the prediction of the thermo-hydraulic performance of Mubea Rollbonded Cooling Plates, 118 historic CFD simulations were used for model training and verification.

As shown in Sect. 3.1 the overall prediction error is below 10%. Given the fact that predictions can be generated in quasi real-time, the model performance is rated well. With the help of such a model the designer can quickly evaluate dozens of design ideas until a reaching certain maturity. Nevertheless, it is necessary to validate the final design with the legacy CFD process and make adjustments were necessary.

Whenever a critical area in the prediction shows a high uncertainty, this might be an indication for the necessity to evaluate the design with a full blow CFD run. With the uncertainty prediction shown in Sect. 3.2, the designer is enabled to judge when a CFD verification is required.

After studying the generalization capability of the algorithm in Sect. 3.3, it can be concluded, that an existing model can be adapted well to new circumstances. In the presented study, the prediction error was below 5% after adding five samples from an unseen project. With Mubea's PLM and SDM databases, the amount of project data is continuously growing. Automated simulation workflows in the hands of CAD-designers further accelerated this data growth.

Using historical data, expert users can now create model training workflows on the NCS platform. With web technology, the finalized models are deployed via REST-APIs or web interfaces. This enables Mubea CAD-designers to evaluate the models in batch mode or to investigate the performance in an interactive manner easily.

As explained in Sect. 1, optimization methods are hard to apply to cooling plate design problems. With the NCS approach, optimization becomes feasible again, by searching for optimal topologies in the GCNN embedding. To achieve this we look forward to integrate more simulation load cases in the training process. This enables the designer to firstly evaluate manufacturability, stiffness and durability more quickly and secondly delivers a holistic model for the topology search.

Data from manufacturing and quality assurance delivers information about the scatter of material properties, in conjunction with manufacturing simulations it is possible to derive the geometric variation of the channel geometry. Training the NCS algorithm with this data will enable us to include 3D manufacturing tolerances in the topology search and guarantee robustness and reliability.

References

1. Bralla, J.G.: Handbook of manufacturing processes. Industrial Press Inc., U.S. (ISBN: 978-0-8311-3179-1)
2. Bay, N., Clemensen, C., Juelstorp, O., Wanheim, T.: Bond strength in cold roll bonding. CIRP Ann. **34**(1), 221–224 (1985)
3. Klinke, N.: The use of HPC in optimal design of TRB lightweight vehicle components. In: 8th european altair technology conference, Paris, France, 29 Sept.–1 Oct. 2015
4. Ryberg, A.-B.: Metamodel-based design optimization—a multidisciplinary approach. Dissertation, Linköping University (2013)
5. Kayvantash, K., Thiam, A.-T., Ryckelynck, D., Chaabane, S. B., Touzeau, J., Ravier, P.: Model reduction techniques for LS-Dyna ALE and crash applications. In: 10th LS-Dyna conference, Würzburg, Germany, 15–17 June 2015

6. Astrid, P.: Reduction of process simulation models: a proper orthogonal decomposition approach. Dissertation, Eindhoven University of Technology (2004)
7. Frenzel, M., Ollar, J., Büttner, C., Finotto, V. C., Fliesser, M.: Automotive crashworthiness optimisation using machine learning to emulate engineering expertise. In: 14th World congress of structural and multidisciplinary optimization, Boulder, USA, 13–18 June 2021
8. Baque, P., Remelli, E., Fleuret, F., Fau, P.: Geodesic convolutional shape optimization. In: Proceedings of the 35th international conference on machine learning. Stockholm, Schweden (2018)

Innovative Module Design with Actice and Passive Cooling of Traction Batteries

David Löffler[1](✉), Rico Schmerler[1], Markus Grünert, Jan Clausen[2], and Simon Schmidt[2]

[1] Department Lightweight Design, Fraunhofer Institute for Machine Tools and Forming Technology IWU, Reichenhainer Strasse 88, 09126, Chemnitz, Germany
{david.loeffler,rico.schmerler}@iwu.fraunhofer.de
[2] Fraunhofer Institute for Manufacturing Technology and Advanced Materials IFAM, Wienerstr. 12, 28359, Bremen, Germany
{jan.clausen,simon.schmidt}@ifam.fraunhofer.de

Abstract. The shift toward electromobility requires innovative solutions for battery design and the associated peripherals in order to increase the efficiency and safety of electric vehicles and increase sustainability. This major challenge was taken up by a consortium of five Fraunhofer institutes as part of a research project. The goal was to develop new designs and manufacturing technologies for e-mobility applications, which would then be implemented in a resource-efficient battery module demonstrator. This module demonstrator (Figure 1) initially consists of two identical cast aluminum half-shells with integrated copper channels for active cooling of the battery cells. These halves are bonded together via fire retardent fiber composite side panels (right, left and back). The adhesive used is thermally releasable, allowing easier disassembly. The module lid is attached to the front face and consists of a metal foam sandwich, which is infiltrated with phase change material (PCM). The PCM serves as a passive cooling system for the power electronics located on the backside of the lid, enabling thermal load peak flattening and increased temperature homogeneity within the power electronics. The heat stored in the PCM can be released to the environment or be removed via the active cooling system subsequent to the peak loads. The overall cooling management of the module thus also allows very heat-intensive charge and discharge cycles. Thin fire protection materials are located between the cells to prevent thermal propagation in the event of a thermal runaway of a cell. Several modules can be interconnected to form a stable unit that can be expanded as required.

Keywords: Traction batteries · Battery module · Lightweight design · Metal foam · Active and passive cooling

K. Dröder and T. Vietor (Eds.): *Future Automotive Production Conference 2022, Zukunftstechnologien für den multifunktionalen Leichtbau*, pp. 219–227, 2023.
https://doi.org/10.1007/978-3-658-39928-3_16

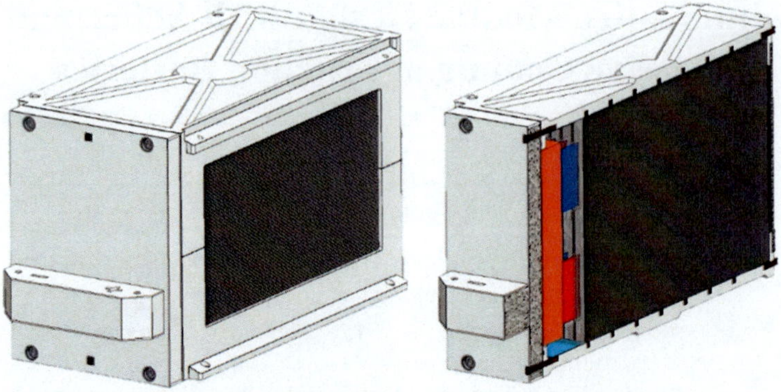

Fig. 1. FutureFlexPro Module complete (l.) and cut (r.). (Source: Fraunhofer IWU)

1 Introduction

The ongoing electrification and digitalization in the automotive sector as well as the intensifying requirements regarding sustainability and resource efficiency pose new challenges for the automotive industry in manufacturing and product development [1]. The social and political relevance of the topic of sustainability continues to grow and is also becoming visible at EU level [2]. The project **futureFlexPro** was therefore launched within the Fraunhofer-Society. In this project, which is funded by the German Federal Ministry of Education and Research (BMBF), the interdisciplinary consortium consisting of five Fraunhofer Institutes (IWU, IAO, IFAM, IST and WKI) and the Fraunhofer Center for Lightweight Construction and Electromobility in Wolfsburg developed demonstrators and concepts of variant-flexible and eco-efficient system components for future vehicle generations in the sense of a holistic circular economy. In the sub-project "battery module construction", a concept for the structural redesign of a hybrid battery module housing was created and manufactured in the form of a demonstrator module. The focus of the conceptual design was on better dismantling capability, improved thermal management and fire protection. This futureFlexPro module is presented and explained in detail below (Fig. 1).

2 The Futureflexpro Module

The vehicle battery systems currently available on the market consist of a battery housing containing individual modules, which in turn contain individual cells. The mechanical functions and the different cooling system are usually located at system level, such as in the battery system with pouch and prism cells of the Mercedes EQS [3, 4]. In the case of round cells, the market leader Tesla relies on a cooling system at module level [5].

The developed module consists of a cell stack built up from twelve individual pouch cells. Two identical cast aluminum halves with fiber composite side panels enclose this cell stack. For self-sufficient use, the module has its own power electronics for monitoring and controlling the cells and can be easily opened for maintenance and repair purposes via the screwed on module lid (Fig. 2). The entire module is 150 mm wide, 184 mm high and 340 mm deep. Mainly the battery cells used inside the module determine its size.

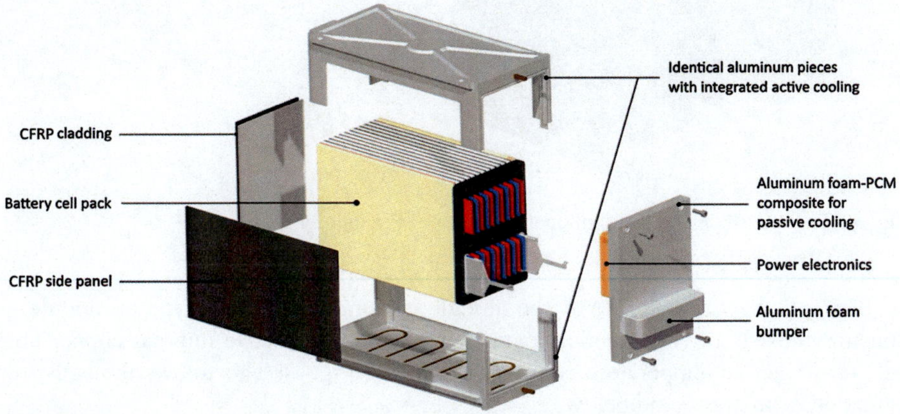

Fig. 2. The futureFlexPro battery module. (Source: Fraunhofer IWU)

The core of the battery module is the cell stack (see Fig. 3). The stack consists of twelve LiFePO4 pouch cells with 3.3 V nominal voltage. These cells have a capacity of 20 Ah each and are connected in series (in 12 s) via a connector component. The voltage of the cell stack is thus 39.6 V. Between the cells there is a high thermal conductive sheet (copper or aluminum) to transport the heat from the inside of the cells to the edge as well as a layer of fire protection material, which also compensates the swelling of the cells by two thicker layers. As shown in Fig. 2 a connection element is attached to each the plus and minus pole for the electrical connection of the stack with other modules or an electrical consumer.

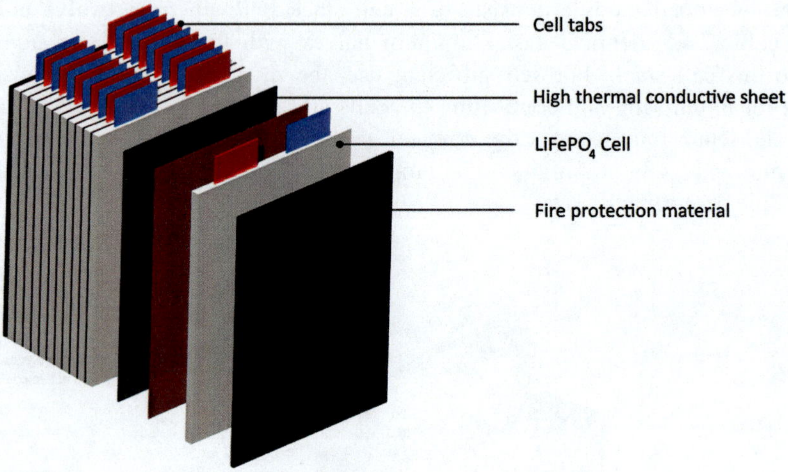

Fig. 3. The cell stack and its components. (Source: Fraunhofer IWU)

The cell stack is enclosed by the module housing. In this new type of module, the housing consists of two identical parts. These half-shells have internal copper channels for an active temperature control system (see Fig. 4). This allows the cells to be tempered from above and below.

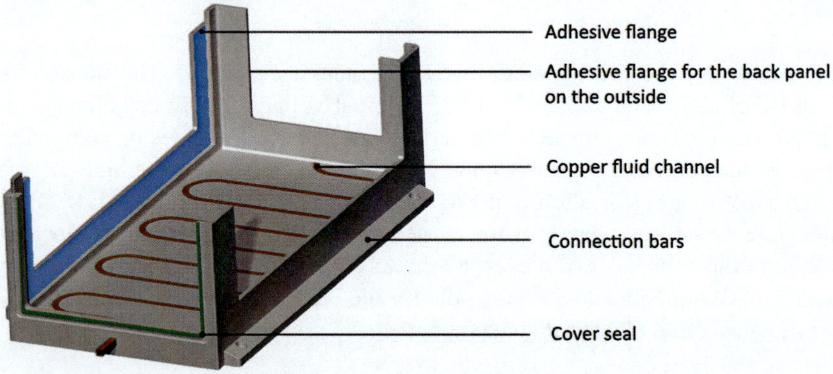

Fig. 4. Module shells with integrated copper fluid channels. (Source: Fraunhofer IWU)

The module halves were designed to be manufactured from aluminum using a low-pressure casting process to integrate the fluid channels. During the production of the castings, the copper fluid channels were inserted into 3D-printed core packages, where they were then enclosed by the molten aluminum.

The two housing shells are joined together via fiber composite (CFRP) side and rear walls with thermally separable adhesive. To assemble the module, the complete cell stack is therefore pre-stressed over the side panels and placed between the module halves. In the next step these are brought together and the stack pretension is released (Fig. 5). A small amount of residual pre-stressing remains on the cells through the module halves to counteract swelling of the cells.

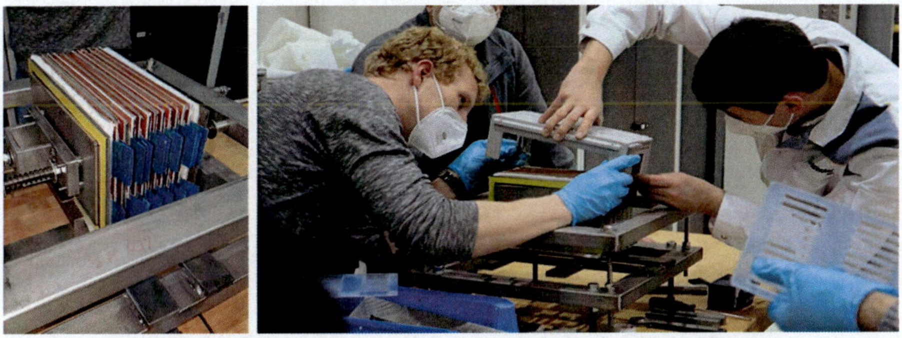

Fig. 5. Assembly of the module: positioning of the cell stack in the joining device (left), assembly of the module shells (right). (Source: Fraunhofer IWU)

The front of the module is tightly closed by a lid with bushings for the cooling lines and electrical connections. With regards to good assembly performance, the lid is fixed with screw connections. On the outside of the lid in the area of the electrical connections a bumper made of an aluminum foam sandwich (AAS) structure is placed, which protects the module from side crash loads. However, the lid also fulfils a further function. The module's power electronics are located on the backside of the cover. These generate a load-dependent energy loss in the form of heat, which results in a temperature increase of the power electronics themselves and the surrounding areas. In the developed module, the lid serves as a passive temperature control system or rather as a buffer storage. It consists of an AAS structure additionally infiltrated with a phase change material (PCM). The sandwich base plate of the front lid containing a closed cell aluminum foam core layer was produced via powder metallurgical foaming route. The complete foam process is shown in Fig. 6. During the foaming process the sandwich aluminum cover sheets build a metallic join with the foam. The base plate setup is AAS15/1/1, whereas the bumper setup an AAS25/2/2. The bumper is joined vertically on the base plate. Inserts for joining the lid with the module frames are placed and the sandwich is sealed along the edges.

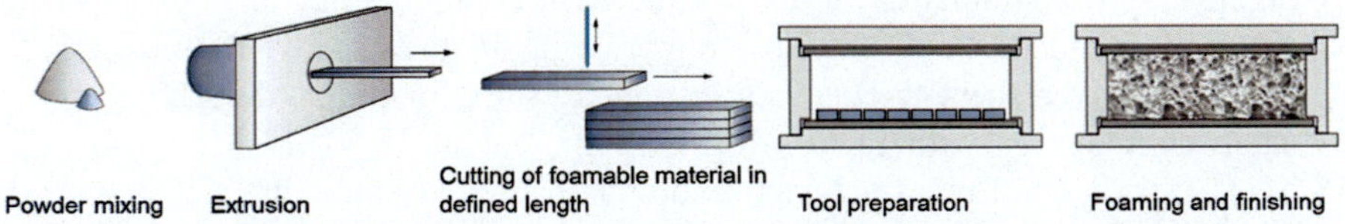

Fig. 6. Powder metallurgical production route for closed cell aluminum foam sandwiches with aluminum cover sheets (AAS). (Source: Fraunhofer IWU)

The by definition closed cell foam, which consist of cell rods as well as cell walls, has micro and macro cracks in its wall structure enabling the foam to be infiltrated with the PCM. For that reason, the foam offers a permeability and enabling it to be infiltrated with fluid. The PCM was infiltrated into the AAS base plate with an infiltration plant and metering unit, see Fig. 7. The infiltration process was pressure supported. Finally, the infiltration ports of the lid were sealed.

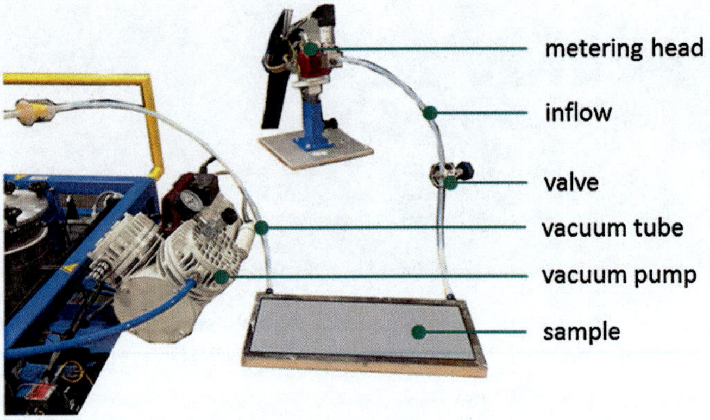

Fig. 7. Infiltration setup for infiltrating AAS with PCM. (Source: Fraunhofer IWU)

The PCM, in this case paraffin, is able to absorb thermal energy, store it and release it again. This is achieved by a phase change from solid to liquid and vice versa. Although PCM has a good heat capacity, it has a low thermal conductivity of 0.2 W/mK. However, the infiltration of the PCM into the aluminum foam compensates this disadvantage, as the aluminum ridges and walls act as a thermal conduction matrix, thus enabling rapid heat transfer into the storage material. The thermal conductivity of the composite material with a density of 0.5 g/cm^3 is 17 W/mK [6, 7].

Paraffin is used in the front lid, as it is particularly suitable for this application due to its corrosion behavior. Paraffin waxes are available with different melting points and can thus be selected to suit the specific application. Thermal simulations were carried out to dimension the AAS-PCM composite in terms of foam cover layer thickness and the selection of a suitable paraffin wax. A fast charge condition of 20–80% State of Charge (SOC), where the power electronics generate 50 W of dissipated heat, was selected as the load boundary condition. The goal was the temperature not exceeding 50 °C to prevent damage to the electronic components.

Figure 8 shows simulation results for different configurations of AAS and PCM. The first number of the AAS description represents the total sandwich thickness, followed by the cover sheet thicknesses in mm. Results for the AAS15/1/1 and AAS 20/1/1 in combination with PCM with melting areas at 35 °C (PCM35) and 44 °C (PCM44) as well as a 1 mm thick aluminium plate reference structure are shown.

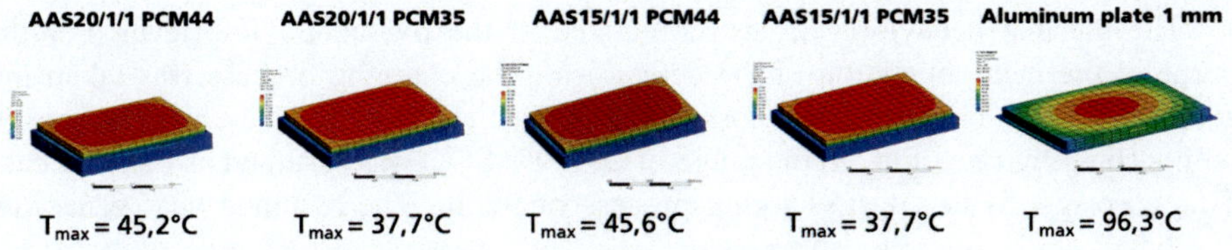

Fig. 8. Thermal simulation results for maximum temperatures (T_{max}) of the front lid with power electronics for different lid setups and PCM. (Source: Fraunhofer IWU)

The simulation showed a significant reduction of 52.6% in the maximum temperature for the PCM setup with a melting point at 44 °C and a reduction of 60.8% for the PCM with a melting point at 35 °C relative to the reference. With both PCM44 and PCM35, the maximum temperature could be kept below the target mark of 50 °C. Furthermore, it can be seen that increasing the foam thickness from 15 mm to 20 mm has only a small influence on the resulting maximum temperature. For the planned application, the configuration AAS15/1/1 with PCM35 offers the biggest potential. This is confirmed by the temperature-time curve shown in Fig. 9.

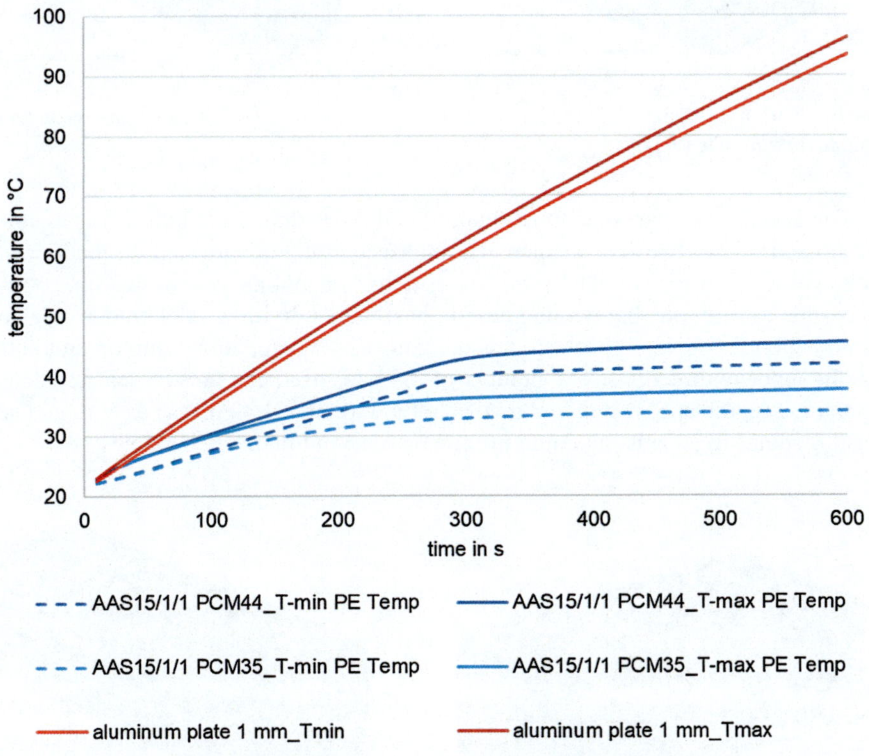

Fig. 9. Temperature-time curve of the maximum and minimum temperatures of the power electronics for different cover configurations. (Source: Fraunhofer IWU)

The thermal behavior was also simulated at the overall module level. For this purpose, the heat generation of the cells during the charging process was taken into account in addition to the heat loss of the power electronics. The charging process with a charging rate of 4C in the range of 20–80% SOC was assumed as the load case. This corresponds to a fast charging process of 9 min. The assumed heat generation was based on the internal cell resistances.

The fluid streams in opposite directions through the two cooling channels of the module shells. The assumed cooling medium is water-glycol with a flow velocity of 18 l/min and the ambient temperature is 20 °C. The system simulation result in Fig. 10 shows that the maximum temperature of the cells does not exceed 38 °C.

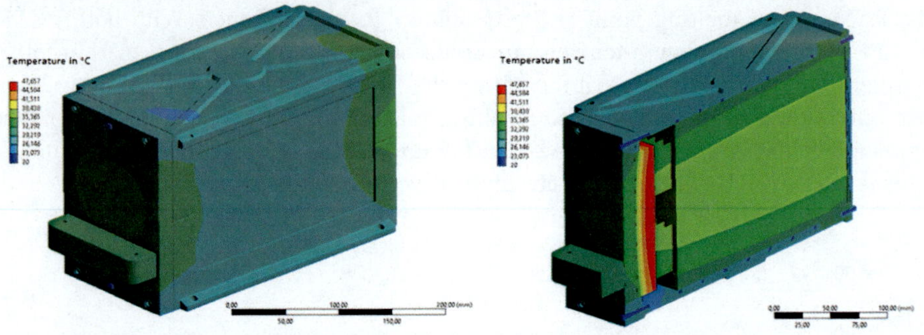

Fig. 10. Thermal simulation of the entire module; full model view (left), sectional view (right). (Source: Fraunhofer IWU)

The scalability is another functionality of the developed module. It is possible to line up and connect several of them via interface structures located on the sides and back, as shown in Fig. 11. The fluid channels of the cooling system line up, so that it is possible to form one big cooling circuit, best with a dividing inlet on one side and a collecting outlet on the opposite side analogous to the simulation with top and bottom flow in opposite directions. The number of modules used can thus be selected according to the specific application and thanks to the modular design, it is easy to add additional modules to an existing configuration or swap out damaged modules.

Fig. 11. Connection of several modules. (Source: Fraunhofer IWU)

3 Conclusion

The battery module demonstrator developed in futureFlexPro shows an approach on how future battery systems for mobility applications can be improved with new features like scalability due to a modular design and the energy efficient passive cooling system in combination with an integrated active cooling system. The focus during module development was not only on new technologies for function integration and lightweight materials, but also on ease of maintenance and disassembly for component recyclability. These goals were achieved through a removable lid and a detachable adhesive connection between the module shells and the fibre composite walls. Product design with repair, disassembly and recycling in mind is key for meeting the sustainability requirements, which will continue to be raised in the future, all the way to a circular economy.

Acknowledgements. The authors gratefully thank the German Federal Ministry of Education and Research (BMBF) for funding this project. Funding code: L1FHG42421.

SPONSORED BY THE

Federal Ministry
of Education
and Research

References

1. Kampker, A., Vallée, D., Schnettler, A.: Elektromobilität: Grundlagen einer Zukunftstechnologie, 2nd edn., pp.15–25. Springer Vieweg, Berlin (2018)
2. European Commission, Mobility and Transport News: New transport proposals tarfet greater efficiency and more sustainable travel, 14. Dec. 2021. https://transport.ec.europa.eu/news/efficient-and-green-mobility-2021-12-14_en. Accessed 17 Feb. 2022
3. Malik, M., Dincer, I., Rosen, M.A.: Review on use of phase change materials in battery thermal management for electric and hybrid electric vehicles. Int. J. Energy Res., **40**(8), 1011–1031 (2016)
4. InsideEVs Homepage: Mercedes-Benz reveals technical specs for EQS electric sedan. https://insideevs.com/news/498761/mercedes-technical-specs-eqs/. Accessed 19. Apr. 2022
5. Quora Homepage: How are the Tesla 18650 cells in the battery pack cooled? https://www.quora.com/How-are-the-Tesla-18650-cells-in-the-battery-pack-cooled. Accessed 19 Apr. 2022
6. Mehling, H., Cabeza, L.F.: Heat and cold storage with PCM. Springer, Berlin (2008)
7. Schmerler, R.: Passive temperature control for EV batteries and power electronics with phase change materials and aluminum foam structures. In: 10th international conference "Thermal management for EV/HEV". Berlin (2022). Accessed 22 Feb 2022

Check for updates

Contribution to the Optimization of Metal-Composite Lightweight Structures in Contextof Digital Linked Development Processes

Fabian Folprecht(✉), Felix Bonn, Daniel Reinhold Haider, Sebastian Spitzer, and Maik Gude

Institute of Lightweight Engineering and Polymer Technology (ILK), Technische Universität Dresden, Holbeinstr. 3, 01307 Dresden, Germany
{fabian.folprecht,felix.bonn, daniel.haider,sebastian.spitzer,maik.gude} @tu-dresden.de

Abstract. Hybrid metal-composite structures offer high potential in lightweight engineering. These hybrid structures bring advantages in strength, stiffness and weight. These hybrid structures are predestined for highly stressed elements such as car body parts. The combination of the different materials leads to a complex process in the design as well as in the manufacturing phase with a multitude of adjustable and interacting parameters. This publication focuses on a deep understanding of the hybrid component to further support the development and digital observed manufacturing process. A stiffness-to-weight improvement for the structure is presented. Potentially relevant parameters are identified from a wide range of geometric variables in each structural element. They were used to set up a parameterized model in SolidWorks and perform a CAD-based parametric FE simulation in ANSYS Workbench. The generated results are used to perform a sensitivity analysis to identify relevant parameters. These key parameters can be readjusted in the design process and provide the basis for a Digital Master in terms of geometry parameters.

Keywords: Hybrid structures · Lightweight design · Development methods · Sensitivity analysis · Key parameters

1 Introduction

Due to their outstanding mechanical properties, fiber reinforced polymers (FRP) are becoming increasingly important in the automotive industry. The combination of FRP with metallic components allows the realization of new hybrid metal-composite structures (MCS), which offer advantages compared to classical solutions in terms of degree of function, design space and weight. On a prototype scale, such hybrid

K. Dröder and T. Vietor (Eds.): *Future Automotive Production Conference 2022,* Zukunftstechnologien für den multifunktionalen Leichtbau, pp. 228–, 2023.
https://doi.org/10.1007/978-3-658-39928-3_17

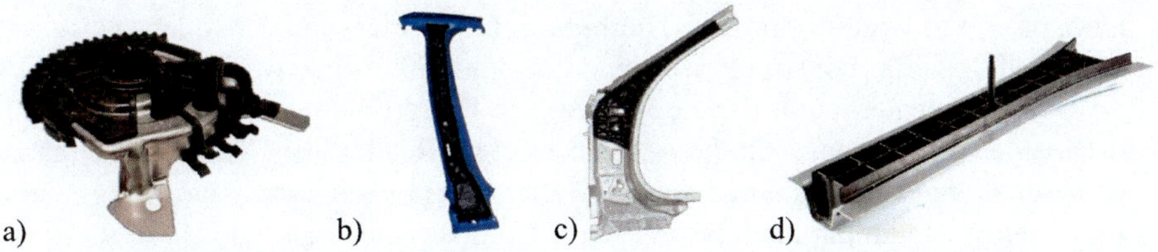

Fig. 1. Hybrid metal-composite structures for automotive application (a–c) and a generic hybrid metal-composite structure (d)

structures have been already designed, manufactured (cf. Fig. 1 a–d) and tested successfully [1–4]. They show a high potential for highly stressed structural elements such as body parts of cars [2]. Furthermore, automated process chains for manufacturing have been demonstrated [4, 5]. Some of the hybrid concepts were successfully transferred into application (cf. Fig. 1c) [3, 4].

The transfer of such MCS into industrial application is still difficult. The multitude of interacting parameters in the areas of design, material, manufacturing and quality assurance leads to a complex and multi-disciplinary development process [6]. To handle such issues in the field of design and dimensioning processes, a variety of methods for structural optimization is known [7]. In general, parameter-based structural optimization can be divided into three classes according to SCHUMACHER: topology optimization, shape optimization and dimensioning. Dimensioning includes the optimization of e.g. the cross-section, wall thicknesses, but also fiber angles or layer build-up. In the case of dimensioning, the FE mesh remains static, since only the properties of the elements need to be adjusted in each iteration, while in the case of shape optimization, a shift of the FE nodes is required.

The relationship between the design parameters and the simulated mechanical properties can be modeled in different ways. In this context, a distinction is also made between mesh-based and CAD-based parameterization of the FE simulation. Mesh-based methods usually need less computing time but lead to incorrect calculation models in case of larger shape variations and require a time-consuming back transformation of the optimization results into a CAD model [8]. For a holistic parametric optimization, it is necessary to handle a complex CAD geometry with a high number of design parameters. Therefore, metamodels are often used as a substitute for computationally intensive simulation when the number of design parameters is high. In contrast to high-fidelity simulations, the processing time of metamodels is significantly lower. Metamodels show an inaccuracy compared to a simulation, but they have a high benefit especially in an early phase of the product development with unknown correlations and interactions of the design parameters.

As an example, for metamodel techniques, optiSLang implements the "model of optimal prediction" (MOP), which performs an automatic feature and metamodel selection via an algorithm in the training phase. The goal in the search for an optimal metamodel is the minimization of the prediction coefficient, a criterion that quantifies the prediction quality of different metamodel variants [9]. A combination of ANSYS and optiSLang is shown by EGERLAND, where the mass and mechanical properties of an automotive fan were optimized. The high number reference dependent parameters and the used skeleton construction showed limitations to link ANSYS and the CAD software for automated parametric optimization. The problem was solved here with a FEM-capable CAD model with the ANSYS DesignModeler [10]. These investigations show the challenges in an automated CAD-based parameter optimization to handle a high number of design parameters.

For the developer, this results in the questions how to define the geometry for the different components and in which component an invested mass has the largest influence. A generic hybrid metal-composite structure (GHS) is developed, which represents a section of a complex and highly stressable area of a car body component (A-pillar) (cf. Fig. 1d) to get a deeper understanding of the MCS. Such a simplified but representative profile in terms of design and technological aspects allows the investigation of the complex interactions between the parameters of geometry, material and technology. The geometry is simplified and symmetrically designed, whereby the respective thicknesses of steel and organic sheet as well as the material combination are taken from the A-pillar. The GHS consists of the four main components hot-formed steel (HFS), cold-formed steel (CFS), organosheet (OS) and an injection molded part (IMP). Each of these components can be described by a various range of geometrical and technological parameters. Due to the integrating design, there is a strong interaction between the parameters of the different components. A correlation study allows to determine the significance of the design parameters to the performance indicators in an early stage of the design process to support the design engineer.

Subsequent, significant parameters can be monitored for quality assurance during production. This parameter monitoring can be used to generate a digital twin of each physical part. According to the WiGeP, the digital twin can be understood as digital representation of a product instance and is generated by the linkage of a digital master and a digital shadow. The digital master as a collection of models contains a geometry of the structure as well as behavioral models. The digital shadow can contain data about an actual structure. This includes operating and status data as well as data on manufacturing. While the product development process brings a digital master as a digital prototype, digital twins can also be prepared for complex use cases like decision support or the controlling of autonomous systems [11].

For a wide industrial application, a better understanding on the geometric influences of the different components in hybrid metal-composite structures is needed. To support this understanding, a geometric design study with a CAD-based parametric FE-Simulation is presented in this publication. Subsequent, parameters that have been identified as relevant can be monitored in the manufacturing and quality assurance process. These data can be used to refeed a digital twin.

2 Identification of Parameters for Digital Linked Processes

The generation of a digital twin of the generic hybrid metal-composite structure (GHS) can be supported by the refeeding of parameters from the manufacturing and quality assurance. To identify the relevant parameters, a workflow for the correlation analysis contributes to the creation of the digital master by corresponding models as well as identifying relevant parameters for capturing in the digital shadow (cf. Fig. 2). In the workflow, the parameters are being identified and systematically reduced to the key parameters. Models of the GHS are being built in parallel that represent different aspects and levels of detail and can thus be brought into the digital master. The first step is to identify the full parameter space. Geometric models are then built, defined

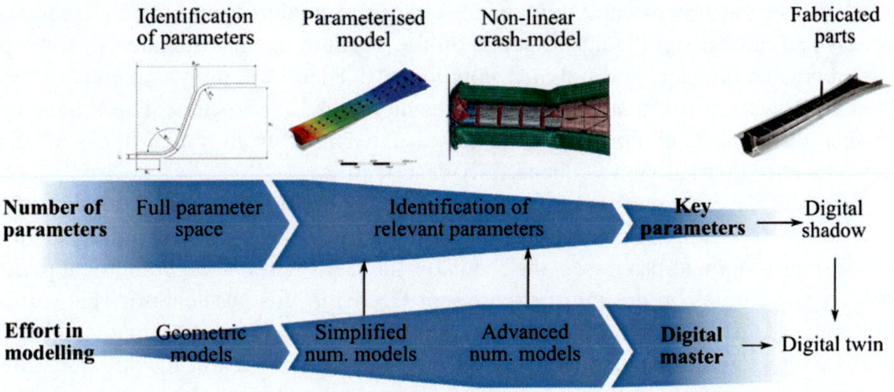

Fig. 2. Identification of key parameters out of full parameter as input for a digital twin (top) and correlating modelling effort (bottom)

according to the identified parameters. Then the parameter space is reduced and the relevant parameters are identified. Simplified models which cover a large parameter space can be used here. With a decreasing number of parameters also advanced numerical models, like non-linear crash models, can be deployed. With their increased modelling effort, they can provide accurate results for a smaller number of parameters. Subsequent to the identification of the key parameters, they can be used for defining the digital twin.

When creating comprehensive models of hybrid structures, a large number of parameters has to be considered. They can be classified into geometric, material and manufacturing parameters (cf. Fig. 3). Geometric parameters mainly include properties related to the dimensions of the component. The manufacturing parameters include settings such as process temperatures or holding times, while the material parameters define the material specification, for example through strengths and stiffnesses. As described in [6], the material and manufacturing parameters can strongly influence each other. This can influence, for example, the strength of the hybrid

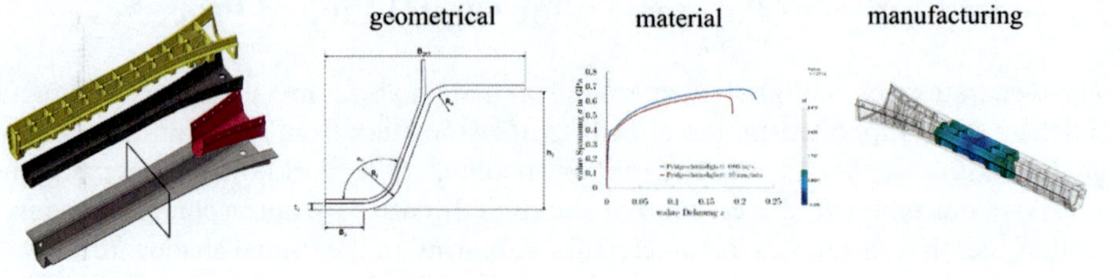

Fig. 3. Individual elements of the metal-composite structure (left) and corresponding geometry, material and manufacturing parameters with a representative visualization

component. Here, process steps such as the pre-treatment of the sheet metal components or also the process and component temperatures during the injection molding process can have a significant influence on the strength against boundary layer separation between the steel and polymer components. Therefore, the parameters of these two categories are summarized below as technological parameters. The interactions between different technological parameters usually have strongly non-linear relationships, some of them are not yet been fully described.

Classical development processes focus the elaboration of the geometry and afterwards allocate required materials and manufacturing processes. According to these classical development processes, the focus for the GHS will be on geometrical parameters for the studies on design of experiment (DOE) in this publication. This component consists of four structural elements (cf. Fig. 3, left), with each individual element having its own geometric parameters. Due to the integral design of the structure, these parameters are directly interconnected, interdependent and interrelated.

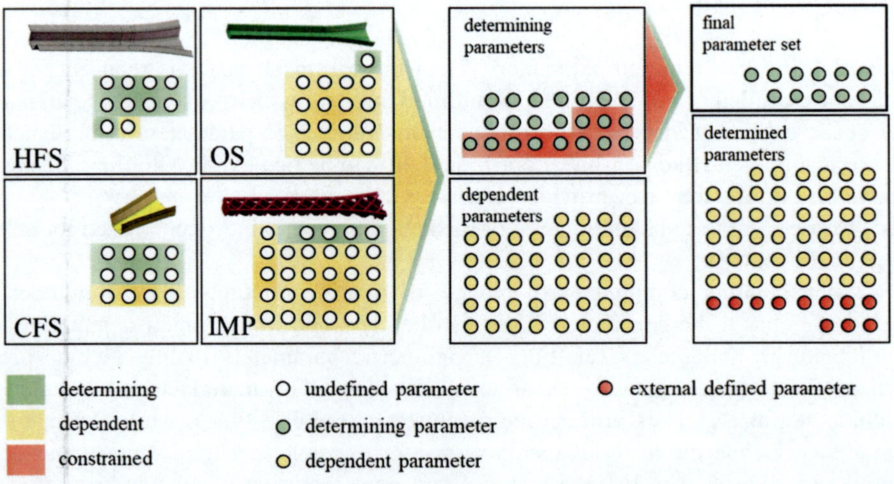

Fig. 4. Stepwise exemplary categorization of the geometric parameters for the components hot-formed steel (HFS), organosheet (OS), cold-formed steel (CFS) and injection molded part (IMP)

3 Digital Linked Model

The respective parameters were identified for the four structural elements in the GHS (cf. Fig. 4). These parameters were categorized in a first step into determining (green) and dependent (yellow) parameters (left). Secondly, the determining parameters can be divided into variable (determining, green) and external defined (constrained, red) parameters (middle), where the external defined parameters result from the boundary conditions, e.g. the total length or width. In the last step (right), the final set of variable parameters can be extracted.

The final parameter set is the foundation for the definition of a parametric CAD-model, whereby the geometry model is controlled by the determining parameters and the determined parameters resulting from given constraints. Other selections on the parameter set can be made based on further constraints as well as estimates of the influence on the overall stiffness. With this method it is possible to extend and reduce the parameter sets to get further insights on the sensitivity.

For this purpose, SolidWorks 2020 (SW) is used. The SW model, as created in this way, can be linked directly (from software to software) with a numerical model

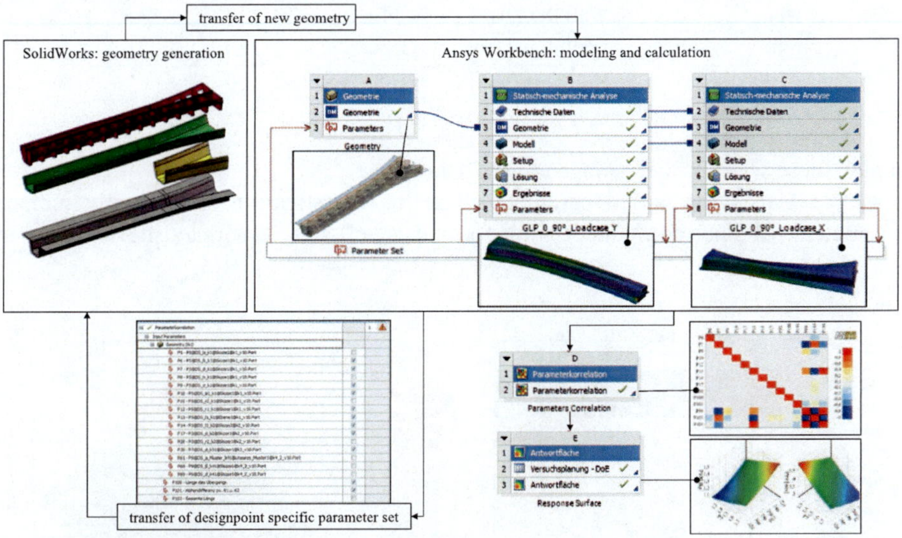

Fig. 5. Interaction of SolidWorks for geometry definition (left) and ANSYS Workbench for FE-modelling and sensitivity analysis (right)

in ANSYS Workbench 2020 R2 (WB), whereby the variable parameters are recognized and can be varied from WB. The interfaces between the two programs are provided and the data formats are compatible. WB controls the geometry modification in SW, which updates the geometry and passes it back to WB for calculation and evaluation. Figure 5 shows an example of the multi-component model from SW (left) with a defined parameter set and the WB project with the parameter linkage (right).

The geometry linked from SW is transferred into an FE model, boundary and load conditions—force in transverse (X) and longitudinal (Y) direction of the profile—are applied and material parameters are defined. The parameter set is created from all variable parameters defined in SW. Based on the parameter set it is possible to use the parameter correlation method of WB. Spearman's rank correlation is selected as the correlation type. In the method of parameter correlation, a DOE is created according to the Latin hypercube sampling. In this case, the recommended size of the experimental design corresponds to 15–20 times the number of input parameters.

Over 50 geometric parameters were identified in the GHS, with 11 shown to be determinant (cf. Fig. 4). To minimize the sample size and robustness of the geometric

Table 1. Table of varied parameters of the correlation analysis

Parameter name	Parameter		Variation range (%)
P7	Wall thickness	HFS	80–120
P10	Floor width	HFS	80–120
P12	Lower radius	HFS	85–115
P17	Wall thickness	CFS	80–120
P38	Wall thickness	OS	80–120
P109	Height difference	HFS/CFS	80–120

model, these were further reduced to six parameters.

Table 1 lists the varied parameters for this investigation and initially includes parameters of the components hot-formed steel (HFS), organosheet (OS) and

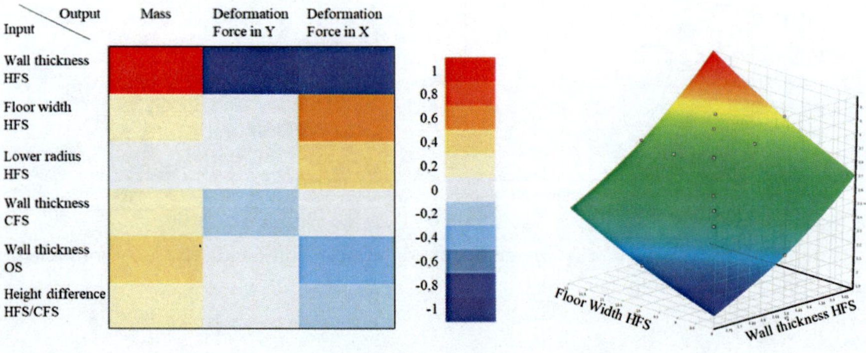

Fig. 6. Correlation matrix for the model (left) and an exemplary response surface for the parameter floor width HFS and wall thickness HFS (right)

cold-formed steel (CFS). The injection molded part (IMP) includes options for topology design in addition to the parameterized geometry. This has an influence on the definition of the mechanical parameters and is difficult to parameterize for a robust model.

The calculation finally results in a correlation matrix, which visualizes the rank of the sensitivity to the selected output parameters. The correlation matrix for the presented example is shown in Fig. 6.

The correlation matrix shows the influences of the individual input parameters on the target parameters mass and deformation of forces in the X and Y directions. The strength of the correlation to the target parameters is color-coded and allows a quick overview of the ranking. The differences between the two load cases are striking. The load case for compression in the Y-direction is dominated by the wall thicknesses of the HFS and OS. For bending in the X-direction, a larger influence is shown by the change in cross-section geometry by means of the width at the bottom of the GHS. The direct influence on the deformation for the second load case is shown for the HFS thickness and the floor width of the HFS in the response surface in Fig. 6 (right). According to the correlation matrix, the increase of the bottom width results in a larger deformation and thus a lower component stiffness. This correlation can be explained by a torsional moment resulting from the load direction, which causes twisting of the section. The unevenly distributed densities and material stiffnesses due to the components lead to an anisotropic structural behavior, especially when loaded outside the plane of symmetry. The superposition of these effects leads to torsion which can explain the high rank of sensitivity for the bottom width. However, whether the influence has to be evaluated as this large for the real component should be considered conclusively e.g. with detailed numerical models.

4 Conclusion

To get a deeper understanding of hybrid metal-composite structures, a parameterized modelling for a GHS is presented. As part of it, a workflow for identifying a parameter space and reducing them to the relevant key parameters is described. In parallel, models are built up to support the definition of a digital twin. In this publication, the workflow for a simplified numerical model is done. In that workflow, the parameters are classified and systematically reduced to a parameter set, which can be investigated in detail. In the next step, a numerical model of the GHS is built using SW and WB for a geometric parameterization.

With the presented method, the sensitivities of the parameters to different target variables for different load cases could be determined with the help of the parameterized model. With the help of a correlation matrix, these sensitivities of all parameters could be visualized. To view individual key parameters, a response surface is used. The results for the presented component show varying sensitivities depending on the load cases. These findings can be used to support developers in the efficient design of hybrid components based on load cases and target variables. Furthermore, the key parameters can be used for quality assurance by measuring them during or after production.

Furthermore, several detailed analyses of the GHS are necessary. The parameter set should be extended to include geometrical parameters previously excluded as well as material parameters or the fiber orientation of organic sheet and injection molded components. Additionally, a research on the topology design of the injection molded part can increase the comprehension depending on the load direction. Afterwards, these results can be transferred to advanced numerical models according to the presented workflow.

Acknowledgements. This research has received funding under the grant number 100339955 ("robust EVP 4.0" project) by the European Regional Development Fund (EFRE) and the German Federal State of Saxony.

References

1. Modler, N., Adam, F., Maaß, J., Kellner, P., Knothe, P., Geuther, M., Irmler, C.: Intrinsic lightweight steel-composite hybrids for structural components. Mater. Sci. Forum **825–826**, 401–408 (2015)
2. Kellner, P.: Zur systematischen Bewertung integrativer Leichtbau-Strukturkonzepte für biegebelastete Crashträger. Dissertation, Technisch Universität Dresden, Dresden (2013)
3. Haider, D.R., Krahl, M., Gude, M., Kellner, P., Knötschke, D.: Quality-assured process chains for the production of highly loaded lightweight structures in metal-FRP design. Plastics in Automotive Engineering, 14.–15.03.2018, Mannheim (2018)
4. Gude, M., et al.: Qualitätsgesicherte Prozesskettenverknüpfung zur Herstellung höchstbelastbarer intrinsischer Metall-FKV-Verbunde in 3D-Hybrid-Bauweise. Plattform FOREL Abschlussbericht Q-Pro (2018)
5. Koshukow, W., Liebsch, A., Kupfer, R., Troschitz, J., Schneider, F., Gude, M.: Entwicklung und Aufbau einer automatisierten Prozesskette für die Herstellung komplexer Kunststoff-Metall-Hybridstrukturen. 27th TECHNOMER, 4.–5.9.2021, Chemnitz (2021)
6. Haider, D.R., et al.: (2021) Contribution to Digital Linked Development, Manufacturing and Quality Assurance Processes for Metal-Composite Lightweight Structures. In: Dröder, K., Vietor, T. (eds.) Technologies for economic and functional lightweight design. Zukunftstechnologien für den multifunktionalen Leichtbau. Springer, Berlin (2021)
7. Schumacher, A.: Optimierung mechanischer Strukturen – Grundlagen und industrielle Anwendungen, 2nd edn. Springer, Berlin (2005)
8. Seiler, M.S.: Geometric constraints in the context of geometry-based structural optimization of machine components with free-form entities. Dissertation, RWTH Aachen University (2013)
9. Gräning, L.: Automation of multi-disciplinary analysis processes with ANSA/META and optiSLang. 8th BEFORE REALITY conference, Munich (2019)
10. Egerland, M., Roos, D., Will, J.: Optimization of a fan shroud by ANSYS/DesignModeler and optiSLang. 25th CADFEM Users' Meeting 2007, Dresden (2007)
11. Stark, R., et al.: WiGeP-Positionspapier zum Thema „Digitaler Zwilling". Zeitschrift für wirtschaftlichen Fabrikbetrieb vol. 115 issue. s1, Hanser, Munich (2020)

Reports from the Research Clusters

Cluster of Excellence Living, Adaptive and Energy-Autonomous Materials Systems (*liv*MatS)

Thomas Speck[1,2](✉), Monika E. Schulz[2], Anna Fischer[2,3], and Jürgen Rühe[2,4]

[1] Plant Biomechanics Group, Faculty of Biology, Botanic Garden, University of Freiburg, Schänzlestrasse 1, 79104 Freiburg, Germany
thomas.speck@biologie.uni-freiburg.de

[2] University of Freiburg, Cluster of Excellence livMatS @ FIT—Freiburg Center for Interactive Materials and Bioinspired Technologies, Georges-Köhler-Allee 105, 79110 Freiburg, Germany
monika.schulz@livmats.uni-freiburg.de,

[3] Institute for Inorganic and Analytical Chemistry (IAAC), University of Freiburg, Albertstr. 21, 79104 Freiburg, Germany
anna.fischer@ac.uni-freiburg.de

[4] Department of Microsystems Engineering (IMTEK), Laboratory for Chemistry & Physics of Interfaces, Albert-Ludwigs-Universität Freiburg, Georges-Köhler-Allee 103, 79110 Freiburg, Germany
ruehe@imtek.uni-freiburg.de

Abstract. The Cluster of Excellence "Living, Adaptive, and Energy-autonomous Materials Systems" (*liv*MatS) develops bioinspired materials systems that adapt autonomously to various environments and harvest clean energy from their surroundings. The intention of these purely technical—yet in a behavioral sense quasi-living—materials systems is to meet the demands of humans with regard to pioneering environmental, sustainability and energy technologies. The societal relevance of autonomous systems and their sustainability thus plays a crucial role in their development within the framework of *liv*MatS. The current contribution provides an overview of the vision, research agenda and research goals of *liv*MatS.

Keywords: Energy Autonomy · Adaptivity · Longevity · Societal Implications

1 Introduction

The vision of the Cluster of Excellence *liv*MatS is to merge the best of two worlds, the biological and the technological realm, to develop living, adaptive and energy-autonomous materials systems. While today's materials are largely static, in that their

K. Dröder and T. Vietor (Eds.): *Future Automotive Production Conference 2022*, Zukunftstechnologien für den multifunktionalen Leichtbau, pp. 239–, 2023.
https://doi.org/10.1007/978-3-658-39928-3_18

properties (except for ageing) do not change with time, *liv*MatS envisions materials systems that are autonomous from an external power supply and can respond to a changing environment by adapting their properties. Such systems, which are inspired by living nature, will harvest the energy they need from the environment and will show complex adaptive behavior via interactions across all length scales, i.e. from the molecular to the macroscopic level [1].

Although these materials systems do not contain biological cells or living subsystems and are not alive in a biological sense (e.g. exhibit no self-reproduction), they are "vital", meaning autonomous, durable, adaptive, programmable, self-regulating, self-repairing and self-protective, so that they can function under adverse conditions and survive limited damage without encountering a system failure [2–4]. The Cluster is also generating several lines of demonstrators that serve as lighthouses and help to keep the *liv*MatS efforts focused [5–7].

*Liv*MatS also addresses important challenging and overarching questions of the societal perception and philosophical implications of materials systems with life-like features. To do so, Cluster research integrates sustainability analyses, psychological research and philosophical reflection of the approaches developed. The convergence of and interaction between disciplines is the essential driver of the Cluster's approach, crossing disciplinary and institutional boundaries and integrating biology, chemistry, soft matter science, process engineering, energy materials research, physics, and microsystems engineering, as well as philosophy, ethics, behavioral science and sustainability research to achieve a holistic view on the topic of living materials systems [3, 8–10].

2 From Bioinspired/Smart Materials to Living Materials Systems

The key goal of *liv*MatS is to make the transition from equilibrium or ("deeply frozen") meta-stable, and thus static, materials to dynamic, life-like, non-equilibrium materials systems. We have identified key principles to make this decisive advance towards "living" materials systems: (see Fig. 1).

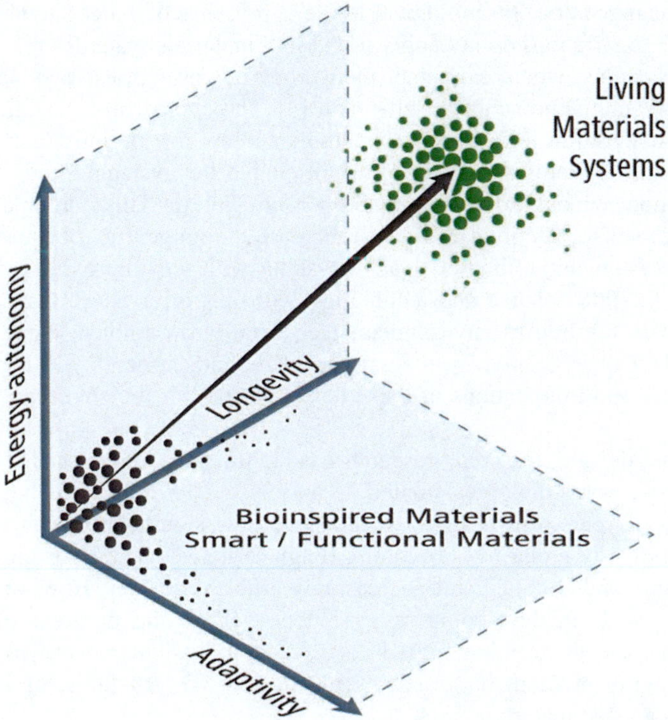

Fig. 1. From bioinspired/smart materials to living materials systems

Energy is a key aspect of this design, as it powers active responses to changes in the environment and enables the performance of (mechanical) work. In the *livMatS* context, energy has to be harvested from the ambient environment and eventually stored to be available on demand. [11] To convert and store the required energy, energy harvesting functionalities must be an integral part of the materials systems to provide true autonomy. Internal control over energy distribution, and active adaption to external signals will require the installation of chemical, structural, and microsystem-based regulatory networks, which will allow for self-regulating properties and generate adaptability [3]. Ultimately, such materials systems may exhibit self-improvement, and capabilities for simple forms of "learning" and training, yet, the materials systems envisioned will allow a (manual) override via human intervention when properties other than those generated automatically are desired.

Such an approach will far surpass current technological pathways to so-called "smart" materials and embedded systems. Our approach will also go well beyond biology. By using the strengths of synthetic and robust materials, applications can be envisioned in environments where biological systems would clearly fail (extreme heat, dryness, pH, etc.). Consequently, such systems will not contain integrated living biological cells or living subsystems, as such cell-based systems would always be limited by conditions mandatory for biological life (e.g. presence of water, moderate temperatures). These re-strictions do not apply to *liv*MatS materials systems.

Having systems, which can adapt their properties in various ways also paves the way for interesting approaches to self-repair [4]. The compartmentalization, miniaturization and integration into complex assemblies allow for the introduction of redundancies into the systems, which in turn will enable the systems to survive (limited) damage without encountering a complete system failure. This combination of fault tolerance and self-protection/-repair will increase the longevity, robustness and resilience of the system and ultimately lead to systems with self-improving properties. The progress of *liv*MatS science and technology will thus offer novel systems that integrate well with the human environment, feed from clean ambient energy, and serve human needs. Consequently, an integral part of *liv*MatS research will be to reflect on the challenges and implications of these developments for the environment and society [8, 9].

In the last decade, materials research has significantly progressed by developing concepts inspired by biological materials and systems which have been optimized over billions of years of evolution. At the same time significant advances have been made in the energy arena by developing (engineering) concepts for energy harvesting, conversion and storage using increasingly efficient devices. However, both fields of science have been developing rather independently, and progress in developing new materials and devices has mostly focused on individual aspects, for example in producing highly efficient harvesters, sophisticated 3D structures, or in generating uniquely responsive materials.

While such impressive advances push forward individual frontiers, the various aspects have only rarely been combined and integrated into a single materials system. This is due to the extraordinary challenge of integrating and harmonizing the often conflicting needs, approaches and concepts, and contradictory operating conditions of the materials and energy worlds, which would be needed for simultaneous integration of energy harvesting, adaptivity and longevity/robustness in one system.

Hence, to reach the goals of the Cluster of Excellence *liv*MatS, it will be essential to overcome this dichotomy and combine efforts and integrate concepts from the various research branches. Obviously such efforts require an interdisciplinary approach with contributions from complementary fields such as biology, chemistry, materials sciences, physics and (microsystems) engineering.

3 Overarching Topics and Research Areas in *Liv*mats

Inspired by living systems, where structure and function are inseparably interwoven on all levels, and in contrast to today's (micro-)systems, where sensors and actors act separately from each other and are driven by an external power supply, *liv*MatS aims for the full integration of structures and functionalities across all interacting hierarchical levels in space and time. The combination and integration of bottom-up approaches (biological or chemical syntheses, nanostructuring, self-assembly) with advanced top-down (micro-)engineering tools (e.g. 2D to 4D printing), will enable flexible, single step assemblies of multi-component and multi-functional materials systems across differing length scales. Although we do not exclude the use of electronics completely, and in some cases employ hybrid approaches, the adaptivity of the systems will be due to intrinsic properties of the materials-system itself and will not be dominated by microchips and motors.

These challenging topics are investigated and combined with each other in four research areas: A—Energy Autonomy, B—Adaptivity, C—Longevity, and D—Sustainability and Societal Implications (see Fig. 2). Research from all four areas feeds into demonstrator projects [1].

Fig. 2. Interplay of the closely interconnected research areas in *liv*MatS (A—Energy Autonomy, B—Adaptivity, C—Longevity, D—Sustainability and Societal Implications). © *liv*MatS/Daniel Hellweg

3.1 Energy Autonomy

The energy for powering autonomous and adaptive functions must be harvested from the envi-ronment, using e.g. solar energy [12], vibrations or thermal gradients [13] as a source. This energy must then be stored for later use in, for example, supercapacitors/batteries [11, 14, 15], energy-rich chemicals [16], or as mechanical stress/strain. To feed the system, this energy frequently needs to be converted into a different form, for example from electrical to chemical energy. Moreover, forms of distribution must be established to either decentralize the power units in the bulk of a material or to distribute energy into remote material areas to be able to energize the systems.

Adaptive reconfiguration of decentralized power units to optimize harvesting is then the next step. As the cluster progresses and the level of system integration increases, the interplay between energy generation and energy consumption required for the system's adaptivity and longevity will tighten, thereby leading to more adaptive and interactive energy harvesting systems with optimized harvesting activities.

3.2 Adaptivity

The harvested energy must be introduced into soft materials to generate active and adaptive materials systems. Such systems will change their properties by adapting to the environment. Adaptivity concepts will be developed which are increasingly directed towards non-equilibrium systems, because active and fast adaptation ideally proceeds from or between kinetically trapped energy-rich states (multistability) or proceeds in dissipative steady states.

In contrast to conventional, responsive (so-called "smart") materials, internal feedback systems and signal processing will empower such systems to adopt self-regulating properties and distinct functional levels in a linear or non-linear fashion (e.g. opposing properties). Such advanced, material inherent information processing will ultimately pave the way for materials systems having a memory, and, as a consequence, capabilities for training, "learning" and self-improvement [3, 6, 7, 17–21].

3.3 Longevity

Just as living entities ensure their survival in a sometimes rather hostile environment, *liv*MatS will develop similar approaches to protect, extend and recover the materials systems' vitality. In this context, the term vitality means that the essential functions of the system persist for a longer period of time even when the system is exposed to hostile conditions. While changes of the outer appearance might occur, functionality is retained. This will be achieved by "training the materials", early damage recognition, and active repair mechanisms as well as the implementation of redundancy that will ensure longevity without system failure [4].

Training here means that if a material comes under stress, i.e. through a mechanical force acting at a certain location, internal transformations occur, which strengthen the material locally. The concept of active self-healing breaks with established strategies of passive self-healing and will require signal-induced activation of recovery mechanisms and/or exploitation of flux states of dissipative materials settings [17–21].

3.4 Societal Implications and Sustainability

As the generation of living materials systems is a revolutionary technology, it will have a major impact on society. Typically, the influence of a disruptive technology is studied only after implementation of the technology has already begun. However, it is not very efficient to fully develop a technology first and only assess whether the developed systems are sustainable or whether they are accepted in society after all these efforts have been made. As can be seen from today's progress in computer sciences, the development of autonomous vehicles or the replacement of humans by expert systems transform society as a whole, while societal discussion lags far behind the technological development process. This discourse will be initiated in the early stages of *liv*MatS and then be ramped up as the technological developments unfold. A key question is how society responds to self-regulating, active, adaptive and autonomous materials with life-like features. *liv*MatS addresses these questions integrating sustainability analyses, psychological research, and philosophical reflection [8, 9].

3.5 Demonstrator Lines

The four different research areas are integrated and focussed in their efforts using technological demonstrators designed and built in *liv*MatS. The demonstrators represent attractive research challenges and serve as lighthouses guiding the research efforts. They will showcase the feasibility of the *liv*MatS technologies developed, be the first steps towards later implementation in collaboration with suitable industrial partners, and will highlight their wide application range.

Demonstrators are being conceptualized along three different lines: two proof-of-concept demonstrators, an artificial venus flytrap [6] and a soft autonomous machine [5], and a set of demonstrators oriented towards practical applications, such as shape adaptive objects [7, 22]. There are myriad potential applications for the materials systems developed in *liv*MatS. One example are "soft" machines that can recognize and grasp objects by feeling them, without the help of a computer. The capability of a materials system to adapt itself to temperatures, lighting conditions, or pressure opens up perspectives in a wide range of application areas, such as protective clothing like helmets and back protectors or prostheses that can adjust themselves to fit the wearer automatically, autonomously and without needing external energy supply—for instance through the use of body heat [23]. Other ideas include packaging materials that grow stronger automatically when placed under stress and building envelopes that level out temperature differences, for example to prevent overheating.

3.6 Institutional Composition, Project Structure and Staffing

*liv*MatS is based at the Freiburg Center for Interactive Materials and Bioinspired Technologies (FIT) and unites researchers from the Faculty of Engineering, the Faculty of Chemistry and Pharmacy, the Faculty of Biology, the Faculty of Mathematics and Physics, the Faculty of Economics and Behavioral Sciences, and the Faculty of Humanities. The cluster's institutional composition reinforces the university's strategic alliance with Freiburg's Fraunhofer Institutes, with the Fraunhofer Institute for Solar Energy Systems (ISE) and the Fraunhofer Institute for Mechanics of Materials (IWM) as partner institutions within the cluster, and is complemented by the Institute for Applied Ecology (Öko-Institut e. V.). The interdisciplinary spokesperson team of *liv*MatS includes Prof. Dr. Jürgen Rühe (Faculty of Engineering), Prof. Dr. Anna Fischer (Faculty of Chemistry and Pharmacy), and Prof. Dr. Thomas Speck (Faculty of Biology).

Since 2019, the cluster has developed and set up three different types of projects to provide flexible formats for the implementation of its research agenda. Doing so has allowed us to strike a balance between long-term projects (36 months) and complementary booster and impulse projects (six to eight and three months respectively). A total of 21 long-term projects, each with a duration of 36 months, were set up following the *liv*MatS calls for projects 2019 and 2020. Long-term projects within the cluster are collaborative, combining the expertise of several Principle Investigators, Responsible Investigators, Postdocs and PhDs within and across the four research areas and the demonstrator area. These long-term projects have been complemented by 25 impulse projects which were completed in 2020 and 2021, and 16 booster projects completed in 2021. This has added to the Cluster's scientific positioning and research output.

Over the past three years, the Cluster has been able to recruit excellent early career researchers for its projects. Currently, a total of 107 researchers at all levels, from doctoral researchers to experienced principal investigators, are working in long-term projects directly funded by the Cluster or associated to it. In addition to the 59 PhD and Postdoc researchers working in these long-term projects, a total of 29 PhD researchers and 14 Postdocs have contributed to *liv*MatS research in booster and impulse projects over the course of 2020 and 2021.

4 Science Communication and Outreach

Science communication and public outreach are of great importance to *liv*MatS [24]. A central component of public outreach is situated at the University of Freiburg Botanic Garden, which is an integral part of *liv*MatS and thus ideally suited to demonstrate similarities and differences in biological and technical materials and illustrate concepts in both the biological and the technical realms. The field of biomimetics is an ideal entry point for discussing the relationship between material properties, material structures and materials systems with the general public who generally have a strong interest in "how nature works", as it directly relates to their daily experience. Most visitors, for example, are interested in how the Venus flytrap functions, why nuts are so very hard, or why pine cones open and close in response to humidity.

In order to facilitate outreach activities, the *liv*MatS Pavilion, a resource-efficient lightweight construction developed through digital planning and robotically manufactured, was built in the Botanic Garden of the University and is now open to the public (see Fig. 3). The pavilion is a model for a bioinspired sustainable construction and stems from the successful collaboration of an interdisciplinary team of architects and engineers from the ITECH master`s program at the Cluster of Excellence "Integrative Computational Design and Construction for Architecture (IntCDC)" at the University of Stuttgart and biologists from *liv*MatS at the University of Freiburg [25].

Fig. 3. Cacti-inspired *liv*MatS Pavilion in the Botanic Garden Freiburg. © ICD/ITKE/IntCDC University of Stuttgart

The inspiration for the *liv*MatS pavilion came from columnar cacti such as the saguaro cactus (*Carnegia gigantea*), and the prickly pear cactus (*Opuntia* sp.), which are characterized by their special wood structure (see Fig. 4). The up to 20 m tall saguaro cactus has – like many other columnar cacti – a cylindrical wooden body that is hollow on the inside and thus particularly light. The individual wooden elements grow together to form a net-like structure, which gives the wooden body additional mechanical stability. The same holds for the cylindrical stem of plants belonging to the genus *Cylindropuntia*. The tissue of the flattened side shoots ("ears") of the prickly pear are traversed by net-like woody fibre bundles, which are arranged in layers and interconnected. As a result, the tissue of the prickly pear is also characterized by a particularly high load-bearing capacity [26–29].

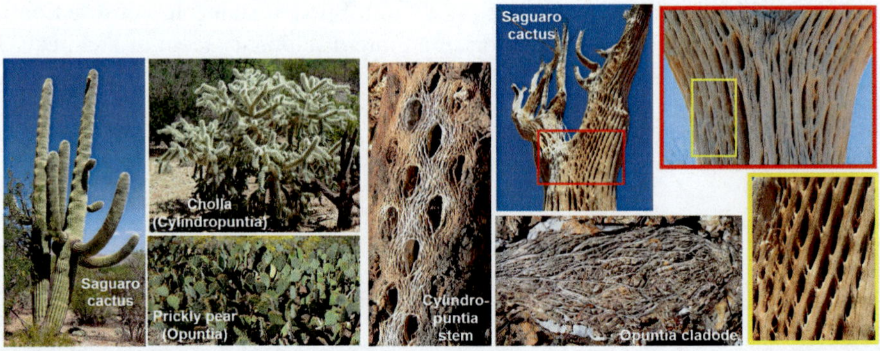

Fig. 4. Habitus and reticulated wood structure of the biological concept generator *Carnegia gigantea, Opuntia* sp. and *Cylindropuntia* sp. © *liv*MatS & Plant Biomechanics Group, University of Freiburg

By analyzing the (micro-)arrangement of the reticulated wood with μCT (Micro-Computer Tomography) and MRI (Magnet Resonance Imaging), and abstracting these net structures (see Fig. 5), the structural and mechanical properties of the cross-linked biological fibre structures could be transferred to the pavilion's lightweight load-bearing elements [28, 29]. As a technically feasible and affordably mode of production robot-assisted coreless winding – replacing the intergrowth that causes the netlike structures in the biological models – was chosen for the production of cross-linked resource-efficient lightweight material systems for architecture.

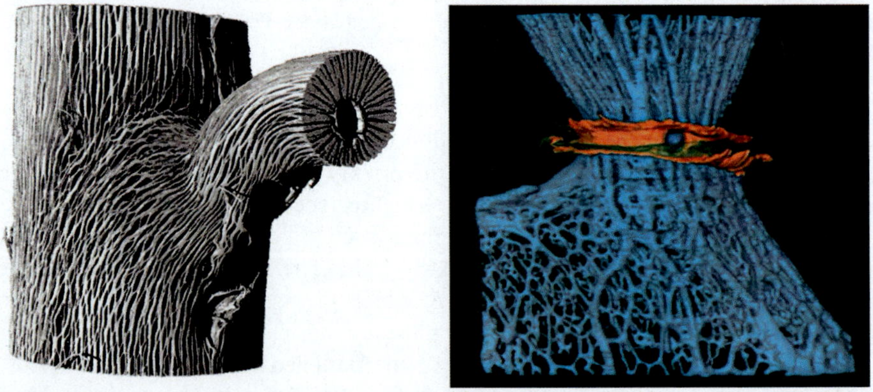

Fig. 5. Micro-Computer Tomography (μCT) scan of the branching regions of a columnar cactus showing the reticulated wood structure (left) and Magnet Resonance (MR) scan of a prickly pear cactus (*Opuntia ficus-indica*) showing the reticulated wood tissue in blue and the periderm in orange (right). © right picture *liv*MatS & Plant Biomechanics Group, University of Freiburg and ILK TU Dresden, left picture from [29].

The supporting structure of the *liv*MatS pavilion consists of 15 flax fibre elements designed by digital planning and prefabricated exclusively from continuous spun natural fibres in a robot-assisted coreless fibre winding process [30–32]. With a total area of 46 m^2, the entire fibre structure weighs only about 1.5 t and is designed to withstand the full snow and wind loads of the applicable building codes (see Fig. 6).

Fig. 6. Resource-efficient lightweight constructions through digital planning and manufacturing by means of winding (core-less winding) "More form – less material". ©ICD/ITKE/IntCDC University of Stuttgart

The pavilion, made (nearly) entirely of recyclable materials, points the way to a new greener architecture of the 21st century. Made of reticulated interconnected robotically wound flax fibers and sisal cords, the structural set-up of supporting elements is inspired by the net-like structure of the wooden body in the trunks of saguaro cacti and by the interconnected wood fibres in the flattened shoots of prickly pear cacti. Contributing to the resource-efficient architectural approach are the load-bearing elements made of renewable raw materials (flax, sisal) and the transparent ceiling elements made of polycarbonate. The ceiling elements can be shredded and remelted at the end of their useful life. The pavilion illustrates how a combination of natural materials used for thousands of years with advanced digital technologies can turn a unique bio-inspired sustainable architecture into reality [30–32].

The pavilion fits naturally into the surroundings of the Botanical Garden in Freiburg. It demonstrates the spatial and structural possibilities of building with natural materials. This creates a distinctive space with a novel architectural expression for visitors to the garden and users of the building. The *liv*MatS Pavilion will serve as an outdoor lecture room for *liv*MatS to vividly communicate the research of the Cluster. Researchers will present their work to the public there, for example, in guided tours or workshops.

5 Conclusion

The work of the Cluster of Excellence *liv*MatS will enable the foundation of a new multidisciplinary center on bioinspired materials research and systems integration with high international visibility at the University of Freiburg. *liv*MatS will perform research across institutional boundaries and help bridge the gap between engineering, natural sciences and behavioral sciences/humanities.

We expect that the progress of research in *liv*MatS will help to meet the needs of a future society that relies heavily on new materials and systems and will provide breakthroughs in the development of materials systems that integrate well into the human environment, sustainably serve human needs, and run on clean energy.

Acknowledgements. The cluster of excellence Living, Adaptive and Energy-autonomous Materials Systems (*liv*MatS) is funded by the Deutsche Forschungsgemeinschaft (DFG, German Research Foundation) under Germany's Excellence Strategy—EXC-2193/1—390951807.

References

1. Cluster of Excellence Living, Adaptive and Energy-autonomous Materials Systems, "Materials Systems of the Future". *Cluster of Excellence Living, Adaptive and Energy-autonomous Materials Systems* (2022) https://www.livmats.uni-freiburg.de/en. Accessed 20 Feb. 2022.
2. Speck, O., Speck, T.: Functional morphology of plants—a key to biomimetic applications. New Phytol. **231**(3), 950–956 (2021). https://doi.org/10.1111/nph.17396
3. Walther, A.: From responsive to adaptive and interactive materials and materials systems: A roadmap. Adv. Mater. **32**(20), 1905111 (2020). https://doi.org/10.1002/adma.201905111
4. Speck, O., Speck, T.: An overview on bioinspired and biomimetic self-repairing materials. Biomimetics **4**(1), 26 (2019). https://dx.doi.org/10.3390%2Fbiomimetics4010026
5. Mazzolai, B., Carpi, F., Suzumori, K., Cianchetti, M., Speck, T., Smoukov, S.K., Burgert, I., Keplinger, T., De Freitas Siqueira, G., Vanneste, F., Goury, O., Duriez, C., Nanayakkara, T., Vanderborght, B., Brancart, J., Terryn, S., Rich, S., Liu, R., Fukuda, K., Someya, T., Calisti, M., Laschi, C., Sun, W., Wang, G., Wen, L., Baines, R., Patiballa, S., Kramer-Bottiglio, R., Rus, D., Fischer, P., Simmel, S.C., Lendlein, A.: Roadmap on soft robotics: Multifunctionality, adaptability and growth without borders. Multifunct. Mater., 100224 (2022) (in press). https://doi.org/10.1088/2399-7532/ac4c95
6. Esser, F.J., Auth, P., Speck, T.: Artificial Venus flytraps: A research review and outlook on their importance for novel bioinspired materials systems. Front. Robot. AI **7**, 75 (2020). https://doi.org/10.3389/frobt.2020.00075
7. Speck, T., Poppinga, S., Speck, O., Tauber, F.: Bio-inspired life-like motile materials systems: Changing the boundaries between living and technical systems in the Anthropocene? Anthropocene Rev. 20530196211039275 (2021). https://doi.org/10.1177%2F20530196211039275
8. Hühn, L., Müller, O., Höfele, P. (eds.): The Role of Nature in the Anthropocene: Defining and Reacting to a New Geological Epoch. The Anthropocene Review (2022) (in press)
9. Möller, M., Höfele, P., Kiesel, A., Speck, O.: Reactions of sciences to the anthropocene: Highlighting inter- and transdisciplinary practices in biomimetics and sustainability research. Elementa Sci. Anthropocene **9**(1), 035 (2021). https://doi.org/10.1525/elementa.2021.035
10. Özdemir, B., Reski, R.: Automated and semi-automated enhancement, segmentation and tracing of cytoskeletal networks in microscopic images: A review. Comput. Struct. Biotechnol. J. **19**, 2106–2120 (2021). https://doi.org/10.1016/j.csbj.2021.04.019

11. Berestok, T., Diestel, C., Ortlieb, N., Buettner, J., Matthews, J., Schulze, P.S.C., Goldschmidt, J.C., Glunz, S.W., Fischer, A.: High-efficiency monolithic photosupercapacitors: smart integration of a perovskite solar cell with a mesoporous carbon double-layer capacitor. RRL Solar **5**(11), 2100662 (2021). https://doi.org/10.1002/solr.202100662
12. Hatt, T., Bartsch, J., Davis, V., Richter, A., Kluska, S., Glunz, S.W., Glatthaar, M., Fischer, A.: Hydrophobic AlOx surfaces by adsorption of a SAM on large areas for application in solar cell metallization patterning. ACS Appl. Mater. Interfaces **13**(4), 5803–5813 (2021). https://doi.org/10.1021/acsami.0c20134
13. Subhash, S.K., Gerach, T., Sherkat, N., Hillebrecht, H., Woias, P., Pelz, U.: Fabrication of μ TEGs based on nano-scale thermoelectric material dispersions. In: Paper presented at the 21st International Conference on Solid-State Sensors, Actuators and Microsystems (Transducers) (2021). https://doi.org/10.1109/Transducers50396.2021.9495526
14. Acker, P., Wössner, J.S., Desmaizieres, G., Esser, B.: Conjugated copolymer design in phenothiazine-based battery materials enables high mass loading electrodes. ACS Sustain. Chem. Eng. **10**(10), 3236–3244 (2022). https://doi.org/10.1021/acssuschemeng.1c07564
15. Esser, B.: Redox polymers as electrode-active materials for batteries. Org. Mater. **1**, 63–70 (2019). http://dx.doi.org/10.1055/s-0039-3401016
16. Beetz, M., Häringer, S., Elsässer, P., Kampmann, J., Sauerland, L., Wolf, F., Günther, M., Fischer, A., Bein, T.: Ultra-thin protective coatings for sustained photoelectrochemical water oxidation with Mo: BiVO4. Adv. Funct. Mater. **31**, 2011210 (2021). https://doi.org/10.1002/adfm.202011210
17. Huber, M. C., Jonas, U., Schiller, S. M. (2022). An autonomous chemically fueled artificial protein muscle. Adv. Intell. Syst., 2100189 (2022). https://doi.org/10.1002/aisy.202100189
18. Kost, J., Bleiziffer, A., Rusitov, D., Rühe, J.: Thermally induced cross-linking of polymers via C, H Insertion Cross-Linking (CHic) under mild conditions. J. Am. Chem. Soc. **143**, 10108–10119 (2021). https://doi.org/10.1021/jacs.1c02133
19. Specht, M., Berwind, M., Eberl, C.: Adaptive wettability of a programmable metasurface. Adv. Eng. Mater. **23**, 2001037 (2020). https://doi.org/10.1002/adem.202001037
20. Mader, M., Hambitzer, L., Schlautmann, P., Jenne, S., Greiner, C., Hirth, F., Helmer, D., Kotz-Helmer, F., Rapp, B.E.: Melt-Extrusion-Based Additive Manufacturing of Transparent Fused Silica Glass. Adv. Sci. **8**(23), 2103180 (2021). https://doi.org/10.1002/advs.202103180.
21. Thimons, L.A., Gujrati, A., Sanner, A., Pastewka, L., Jacobs, T.D.B.: Hard-material Adhesion: Which Scales of Roughness Matter? Exp. Mech. **61**, 1109–1120 (2021). https://doi.org/10.1007/s11340-021-00733-6
22. Tahouni, Y., Krüger, F., Poppinga, S., Wood, D., Pfaff, M., Rühe, J., Speck, T., Menges, A.: Programming sequential motion steps in 4D-printed hygromorphs by architected mesostructure and differential hygro-responsiveness. Bioinspiration Biomim. **16**(5), 055002 (2021). https://doi.org/10.1088/1748-3190/ac0c8e
23. Cheng, T., Tahouni, Y., Wood, D., Thielen, M., Poppinga, S., Buchholz, L., Steinberg, T., Menges, A., Speck, T.: Bio-inspired motion mechanisms: computational design and 4D-printing of self-adjusting wearable systems. Adv. Sci. **8**(13), 2100411 (2021). https://doi.org/10.1002/advs.202100411
24. Speck, O., Speck, T.: Biomimetics and education in Europe: challenge. Oppor. Var. Biomim. **6**(3), 49 (2021). https://doi.org/10.3390/biomimetics6030049
25. Cluster of Excellence Living, Adaptive and Energy-autonomous Materials Systems. Robotically Wound Natural Fibre Construction", Cluster of Excellence Living, Adaptive and Energy-autonomous Materials Systems (2021). https://www.livmats.uni-freiburg.de/en/news-press/robotisch-gewickeltes-naturfasergebaude. Accessed: 20 Feb. 2022.

26. Schwager, H., Haushahn, T., Neinhuis, C., Speck, T., Masselter, T.: Principles of branching morphology and anatomy in arborescent monocotyledons and columnar cacti as concept generators for branched fibre-reinforced composites. Advanced Engineering Materials 12(12), B659–B698 (2010). https://doi.org/10.1002/adem.201080057.
27. Schwager, H., Masselter, T., Speck, T., Neinhuis, C.: Functional morphology and biomechanics of branch-stem junctions in columnar cacti. Proceedings of the Royal Society B **280**, 0132244 (2013). https://doi.org/10.1098/rspb.2013.2244.
28. Masselter, T., Hesse, L., Böhm, H., Gruhl, A., Schwager, H., Leupold, J., Gude, M., Milwich, M., Neinhuis, C., Speck, T.: Biomimetic optimisation of branched fibre-reinforced composites in engineering by detailed analyses of biological concept generators. Bioinspiration & Biomimetics **11**(5), 055005 (2016). https://doi.org/10.1088/1748-3190/11/5/055005.
29. Mylo, M., Hesse, L., Masselter, T., Leupold, J., Drozella, K., Speck, T., Speck, O.: Morphology and anatomy of branch-branch junctions in Opuntia ficus-indica and Cylindropuntia bigelovii: A comparative study supported by mechanical tissue quantification. Plants **10**(11), 2313 (2021). https://doi.org/10.3390/plants10112313.
30. Menges A., Knippers, J. (Eds.).: *Architecture Research Building: ICD/ITKE 2010-2020.* Basel: Birkhäuser Verlag (2020)
31. Bodea, S., Mindermann, P., Gresser, G.T., Menges, A.: Additive Manufacturing of Large Coreless Filament Wound Composite Elements for Building Construction. 3D Printing and Additive Manufacturing, Ahead of Print, 2021. https://doi.org/10.1089/3dp.2020.0346.
32. Gil Pérez, M., Zechmeister, C., Kannenberg, F., Mindermann, P., Balangé, L., Guo, Y., Hügle, S., Gienger, A., Forster, D., Bischoff, M., Tarín, C., Middenhoff, P., Schwieger, V., Gresse, G.T., Menges, A., Knippers, J.: Computational co-design framework for coreless wound fibre–polymer composite structures. J. Comput. Des. Eng., **9**(2), 310–329 (2022). https://doi.org/10.1093/jcde/qwab081.

Potentials and Design of a Virtual Production System for Intelligent Battery Cell Manufacturing

Kamal Husseini(✉), Hans Thomas Augspurger Hernández, Dominik Mayer, and Jürgen Fleischer

Institute of Production Science (Wbk) of Karlsruhe Institute of Technology, Kaiserstraße 12, 76131 Karlsruhe, Germany
{kamal.husseini,hans.hernandez,dominik.mayer2,juergen.fleischer}@kit.edu

Abstract. The increasing worldwide demand for lithium-ion batteries is no longer an estimated forecast, but a fact. Observable trends, such as the increasing variety of battery cell formats and materials, present enormous challenges for the design of production processes. Many cause-effect relationships can be seen in the individual manufacturing processes, but are not yet properly or only partially understood. To meet these requirements and challenges and to ensure effective manufacturing, intelligent processes are needed in battery cell production. Methods of digitization, artificial intelligence and the use of digital twins offer a high potential to optimize the processes both in commissioning and in operation. These methods can be applied within a virtual production system for process optimization in battery cell manufacturing. In this paper, the potential of a virtual production system for battery cell manufacturing is discussed. Further, the design of a suitable infrastructure consisting of standardized interfaces, data models and process models is provided. The result is a system that offers the possibility to virtually quantify cause-effect relationships, to test optimization approaches along the entire process chain of battery cell production and to apply recommendations for action to the real production system. The process step of cell assembly is considered as an example. Here, the simulation model and the associated data model are specified.

Keywords: Battery cell production · Digital twin · Smart production · Virtual production system

1 Introduction

The measures to meet the Paris Climate Agreement of 2015 are reflected in the request for battery cells. The trend of this rising global demand has been confirmed in recent years. Especially worth mentioning is the momentum with which growth has developed. The reasons for this are mainly the increasing demand and application in

K. Dröder and T. Vietor (Eds.): *Future Automotive Production Conference 2022, Zukunftstechnologien für den multifunktionalen Leichtbau*, pp. 253–, 2023.
https://doi.org/10.1007/978-3-658-39928-3_19

electromobility [1]. To meet this demand, production will have to be ramped up in the upcoming years, and the battery cells' performance needs to be improved. This undertaking will face several challenges. Battery production is a complex process consisting of many steps and parameters which significantly influence the product quality. However, many interactions between the process and the product are not yet, or only partially understood [2]. In [3], another challenge is named. Improvements in materials and production in the future will lead to the current battery cell configurations not being permanent. The trend here is towards a greater variety of formats, new materials, continuously thinner electrode foils, and smaller separator thicknesses [4–6]. Consequently, the ramped-up production in the future must be able to react flexibly to changing boundary conditions in battery cell production.

Methods of Industry 4.0 and digitalization offer a possibility to meet the challenges and achieve a more efficient and flexible production process. A digital twin of the production process makes it possible to test the processes without risk in a virtual space, develop optimization approaches, and give recommendations for action to the real production system. First applications of digital twins of lithium-ion batteries and entire production systems as a virtual production system have already been carried out and will be described in the following. In [7, 8], digital twins of lithium polymer batteries are introduced to estimate the state of charge and state of health. Particularly in [9], a reduced-order model is used to decrease computation time and enable real-time applications. An approach using a cloud application is described in [10]. Measured data sets are sent to a cloud platform and are evaluated by a digital twin of a battery management system. In [11], a digital twin of a production system in the automotive industry is proposed. This can be used to optimize the process before an upcoming introduction of a new product variant. In [12], a multiscale simulation approach for production systems in battery cell manufacturing is presented. This approach also considers the dependencies of the product quality over the entire process chain. In [13, 14], the development of a digital twin and modeling methods for the process step of calendering is described. Approaches to simulate the process of single-sheet stack formation are shown in [15]. Further explanations on the development and application of digital twins of energy storage systems are given in [16–18]. In [19], the development of a digital twin of the production machine for flexible cell stack formation is described. The model presented here will be taken up further in the context of this paper.

However, to meet future challenges in battery cell manufacturing, a digital twin is needed that models the production processes, machines, systems and product quality. The following describes the development of such a virtual production system for the cross-process optimization of battery cell production. Thereby, individual aspects of the system are highlighted and the process step of cell assembly is described as a use case.

2 Structure and Development of a Virtual Production System

This chapter describes the structure and design of the virtual production system and its integration into a suitable infrastructure to optimize battery cell production along the entire process chain. The objective is not only to develop virtual process models, but also to provide a corresponding infrastructure to use these models in their entirety. Specific aspects of this system are explained in more detail based on the assembly process of pouch cells. Figure 1 illustrates the overall structure and integration of the virtual production system employing selected processes of battery cell manufacturing

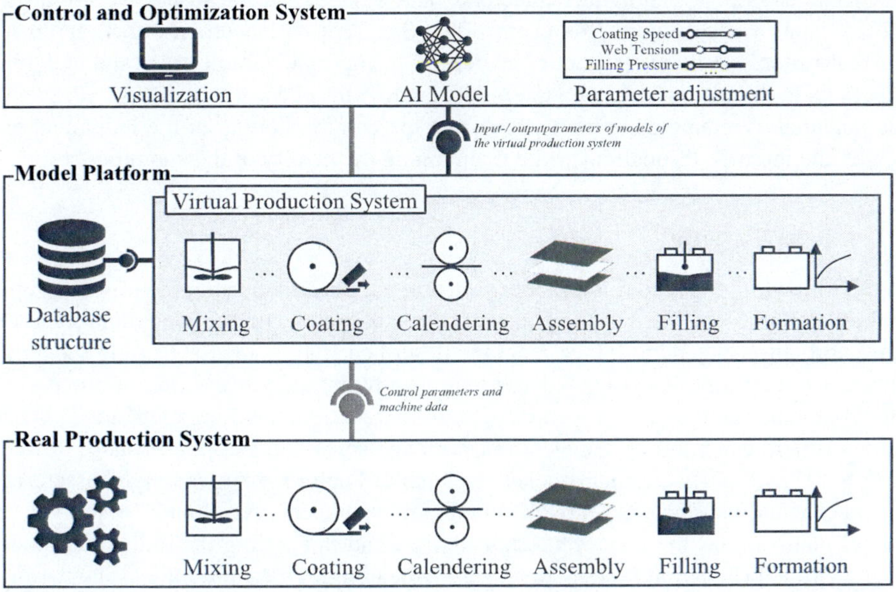

Fig. 1. Design and structure of the virtual production system based on the ViPro project (grant number: 03XP0324C) of the InZePro Cluster

The overall structure consists of three levels. These are the higher level control and optimization system, the model platform, and the real production system. The model platform is the central element of this entire system. Here, models of the individual processes are integrated both as semantic data models and as simulation/prediction models. These models simulate the individual processes and production machines and are intended to represent the real production system in the virtual environment. The real production system includes the machines and equipment and its individual control systems. The control and optimization system is able to request, process, and also send data and information of the different process models of the virtual production system and the real system. The levels are linked to each other via corresponding data interfaces. One way to accomplish this data transfer with a suitable middleware is described in [20, 21]. Here, the implementation of a manufacturing service bus (MSB) and the use of the corresponding cloud-based platform virtual fort knox (VFK) is explained.

The functionality of the virtual production system and its infrastructure is described as the following. The control and optimization system systematically addresses the individual process models of the virtual production system in the model platform. Here, input data such as material characteristics, production volume and required cell format are processed by the individual process models. Output data, such as information on the respective intermediate product properties, machine behaviour and the associated control parameters of the machines and plants are then evaluated, for example, by a machine learning algorithm. Thus, the objective is to determine optimal control parameters of the individual machines and systems for the respective production requirements virtually and without risk. These control parameters are then applied to the real systems. A manual, iterative search for the optimal parameters of the individual processes in the real production system level would thus be avoided. Above all, there is the potential to reduce material waste during the commissioning of the individual processes and increase throughput while maintaining the quality of the end product.

2.1 Model Platform

In the following, aspects of the model platform are described using the process step of cell assembly as an illustration. The focus here is specifically on singulation and cell stack formation. The Coil2Stack machine described in [22, 23] for flexible separation and cell stack formation is considered here as a reference process. In the process, the electrode material is fed over a roller system to a flexible handling element. This handling element can wind up the electrode web and separate electrodesheets of different lengths. These sheets are then stacked on a table. The process is mainly characterized by mechanical aspects, respectively kinematic sequences. As already described, the model platform has the task of structuring data and integrating the individual models of the virtual production system. To ensure a meaningful data structure, a standardized semantic description of the processes is required here. Figure 2 illustrates the detailed structure of the model platform using the example of the cell assembly process step.

The simulation/ prediction model of the cell assembly process is shown on the right side. In this case it is a system simulation by using the software Simcenter Amesim (detailed description in Sect. 2.3). The simulation/ prediction model is configured with appropriate interfaces for the input and output parameters. The provision into

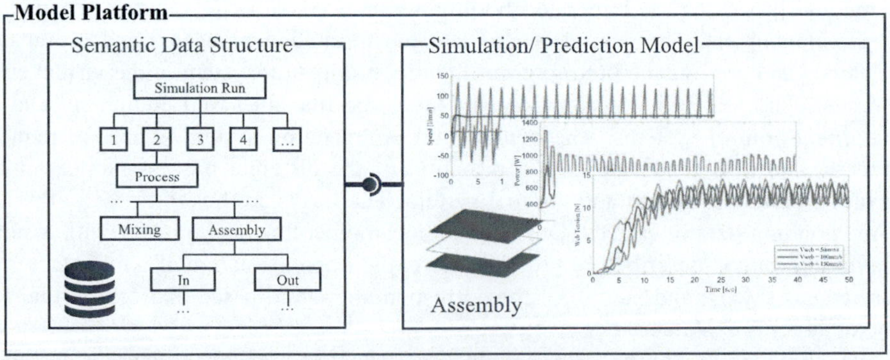

Fig. 2. Detailed illustration of the model platform for the process step of cell assembly

the platform is done by transferring the model into the Functional Mock-up Interface (FMI) standard [24]. The simulation/ prediction model now generates data describing the process and machine operation. For example, the energy demand and the material load are simulated here. On the left side, for each process the corresponding standardized semantic data structure is implemented. Different simulation runs with different input parameters for the simulation/ prediction models are executed. The models' input and output values are structured and stored here. On this basis, the control and optimization system can access simulated datasets along the entire process chain.

Figure 3 shows the semantic data structure for the assembly process model with its in- and output parameters in detail. The input parameters are structured into boundary conditions/general parameters and control parameters. The boundary conditions/ general parameters mainly comprise material-side parameters and specifications for the cell format and throughput to be produced. The control parameters include the parameters that can be set on the production machines. The output parameters are classified into general parameters and quality-related parameters. The general output parameters include variables that each process step exhibits, such as the energy requirement of the machine and the processing time. The quality-related parameters contain values that describe the process quality respectively the quality of the intermediate product. For example, statements on the material stress within the machine and tolerances of the dimensional accuracy of the electrode sheets can be derived from the web tension. The parameters are specified regarding their unit and data type. This data structure is created for each process. The selection of the parameters here requires a deep understanding of the process. However, to be able to perform optimization approaches along the process chain, it is of significant importance that output parameters of one process are equal to input parameters of another process. In this case, the modeled areal weight of the coating after the coating process and the coating thickness after the calendering process are fed into the assembly model.

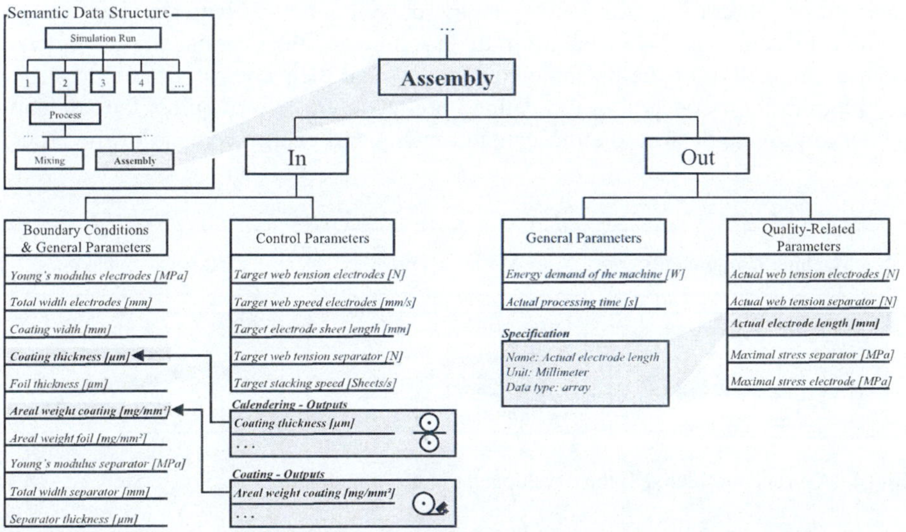

Fig. 3. Data structure for cell assembly with the focus on singulation and stack formation

2.2 Control and Optimization System

The control and optimization system provides selected input parameters to the individual models. Concerning the output data, subsequent interaction relationships are determined across processes and optimal control parameters for the individual process steps are derived for each production scenario and transferred to the real system. The optimization is carried out against the background of different production criteria. These include aspects of quality, throughput and energy demand. In this case, machine learning methods could be used to optimize the process. The data generated in the virtual production system can be evaluated against the background of the production criteria and parameters of the individual process steps can be determined for an overall optimization of the process chain. Furthermore, the control and optimization system includes the Human Machine Interface (HMI) and the visualization of data of the virtual and real production system.

2.3 Process Model of Cell Assembly

In the following, aspects of the actual simulation/ prediction model of the cell assembly process step, focusing on singulation and cell stack formation, are given. The simulation model is built with the software Simcenter Amesim. Individual aspects are based on physical relationships described by differential equations. The modeling of the machine behavior and the determination of the resulting material behavior and intermediate product properties are addressed.

Figure 4 illustrates the aspects considered in the model development. These are the description of the individual machine components, the motion control and the material behaviour. These include, for example, roller systems, screw drives and belt drives. The descriptions are based on equations of motion and friction models. In addition, the control of the machine via corresponding cam plates of the drives is considered. Here, the cam plates defined in the real control system are used. The characteristics of the drives with corresponding delay or inertia elements is as well part of the considerations. The behavior in the machine and the material stress are derived from the modeling of the machine components and their control via the drive models. Optimization concerning the defined production criteria requires the simulation model to provide data on the throughput, quality and energy demand of the process.

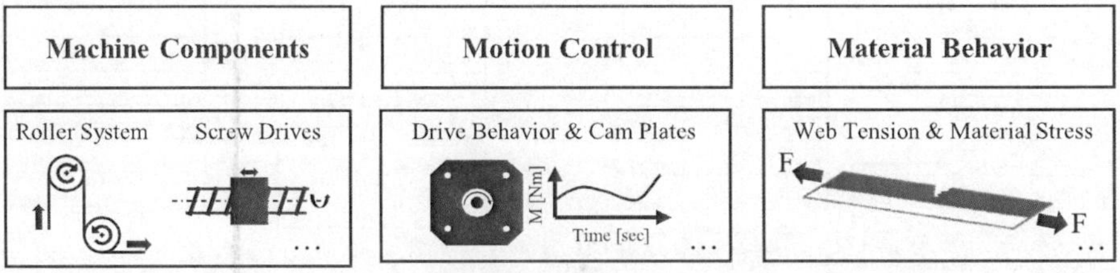

Fig. 4. Aspects considered in the development of the simulation/ prediction model

Here, the throughput is set by web speed, statements on quality are derived from the web tension that occurs, and statements on energy consumption are derived from the power of the electrical drives.

Figure 5 shows the simplified machine concept. The system consists of an unwinder, roller system and handling system. Three drives move the entire system. The unwinder and the handling system are each actuated by a drive for the rotational movement. A vertical movement of the entire handling system is realized by means of a screw drive and a third motor. The position of the web tension measurement is also shown. It is located in front of the handling system. Hereafter, modeled results for singulation of the anode material are presented.

Table 1 illustrates the selected input parameters. Three simulation runs are performed with different target web velocities of the electrode. The descriptions of the material behavior are based on the law of conservation of mass. The corresponding differential equations in the context of web guiding are described in [25].

Figure 6 shows the behavior of the three drives. The modeled speed of the servo motors integrated into the machine is illustrated here. The specified target web speed considered here is 100 mm/s. This motion behaviour drives the entire system. All machine components and the material response are finally affected by this. So also the occurring web tension and energy demand.

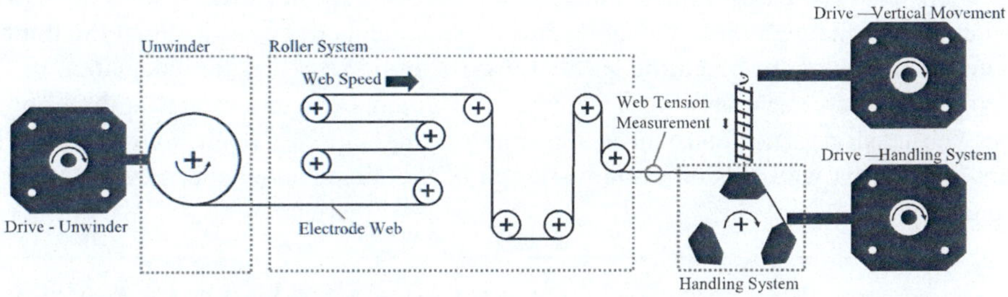

Fig. 5. Simplified illustration of the machine concept

Table 1. Input parameters

Boundary Conditions and General Parameters		Control Parameters	
Young's Modulus Electrode	100 GPa	Target Web Tension Electrode	12 N
Total Width Electrode	200 mm	Target Web Speed Electrode	50; 100; 150 mm/s
Coating Width	147 mm	Target Electrode Sheet Length	210 mm
Coating Thickness	108 µm		
Foil Thickness	11 µm		
Areal Weight Coating	0.236 mg/mm^2		
Areal Weight Foil	0.098 mg/mm^2		

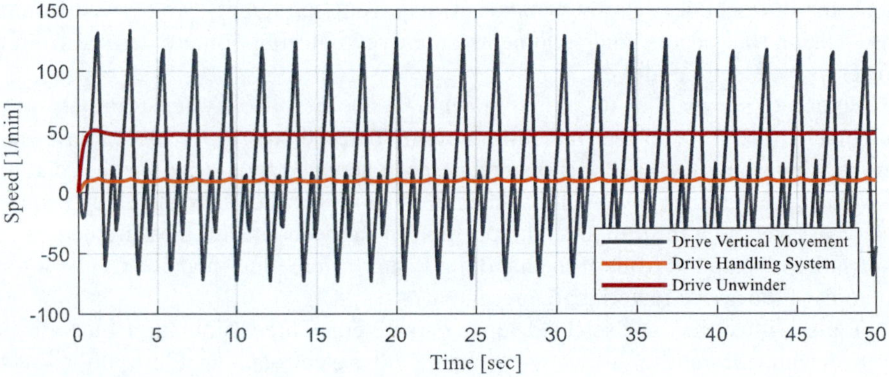

Fig. 6. Modeled speed of the drives for 100 mm/s of web speed

Figure 7 shows an example of the modeled web tension at the measurement position in the machine for different web speeds. It can be seen that there are deviations from the set target web tension of 12 N in all setups. Periodic progressions of the web tension can be recognized. It should be noted that the frequency of these periodic progressions also increases with the increase of the web speed in the system. However, the course of the web tension enables further statements to be made about the dimensional accuracy of the electrode sheets [26]. It can be seen that the fluctuation of the web tension can negatively influence the web guidance in the system. Furthermore, the web tension is not constant at the time of singulation. Conclusions can thus be drawn from the web tension to the variation of the dimensional accuracy of the electrode sheets.

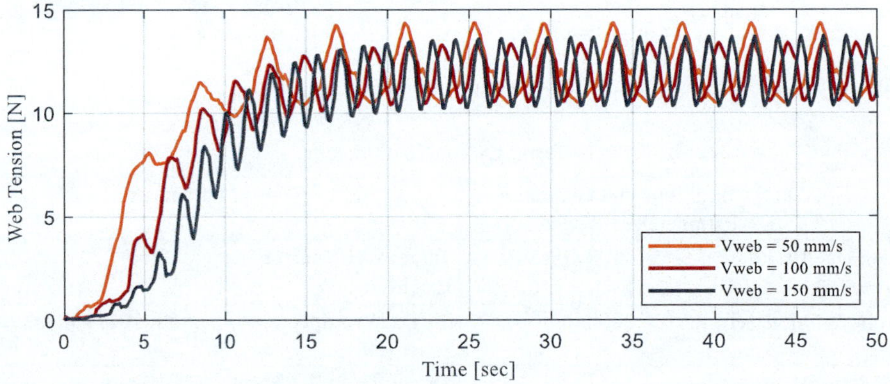

Fig. 7. Modeled web tension at different web speeds

Figure 8 shows the demandmodeled power of the machine at different web speeds. Here it can be seen that the power also increases with the rising web speed. The fluctuations that can be detected correlate essentially with the drive for the vertical movement.

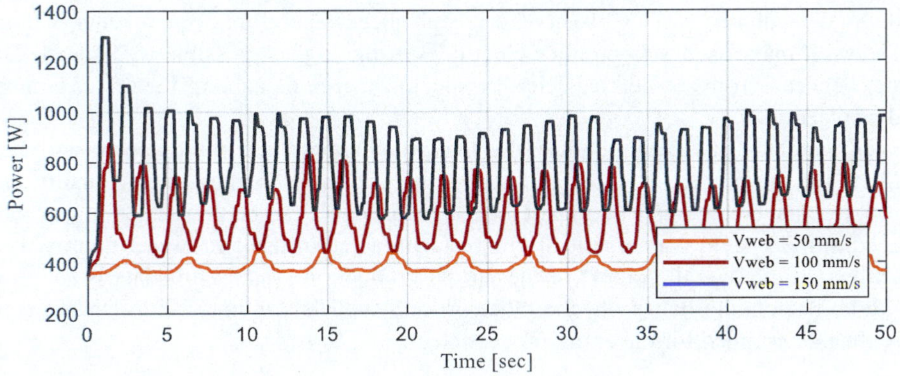

Fig. 8. Modeled power at different web speeds

Figure 9 shows a summary of the modeled results. The quantities were related to a reference battery cell of the Karlsruhe Institute of Technology with a cell capacity C of 20 Ah. The calculation of the throughput is shown in equation (1). Here *LSheet* (= 210mm) donates the length of the sheets, and *NSheets* (= 41), the number of sheets within a cell. The calculation of the specific energy demand is shown in equation (2).

$$Throughput = \frac{v_{web} - C}{L_{Sheet} - N_{Sheets}} \qquad ((1))$$

$$Specific\ Energy\ Demand = \frac{\int_{t1=0sec}^{t2=50\ sec} Power\ (t) \cdot dt}{Throughput - t_2} \qquad ((2))$$

The standard deviation (STD) of the simulated web tensions are illustrated as well. It can be seen that, in addition to the increase in throughput with increased web speed, the standard deviation of the web tension trises too. It can be assumed that at higher web speeds, the electrode sheets are separated with a higher variation of web tension. Therefore the deviation of the dimensional accuracy of the electrode sheets increases with an increase in the web speed. Furthermore, it can be observed that the specific

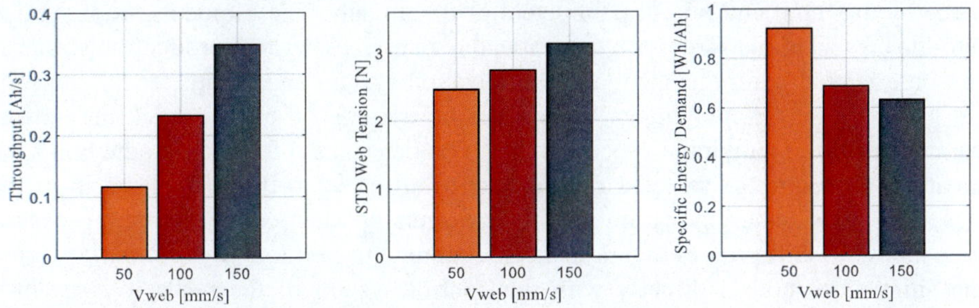

Fig. 9. Results modeled in summary

energy demand decreases with increasing web speed. Therefore, from an energy point of view, it makes sense to produce electrodes at the highest possible web speed. The ratio between produced cell capacity the specific energy demand of the drives is more advantageous.

In summary, this example illustrates the virtual exploration of parameters and their effects for the process stem of cell assembly. Data sets on throughput, quality and energy demand can thus be compared and contrasted. It can be seen that a higher web speed in the system is energetically more favourable but leads to higher quality fluctuations. In the overall context of the virtual production system, this data is analysed with corresponding data from the other process models. Optimal control parameters are then determined for the entire process chain.

3 Conclusion and Outlook

Both the strong global increase in demand for lithium-ion battery cells and the associated trends in material and format diversity present extreme challenges for battery cell production processes. The effects and interactions occurring in the sub-processes and along the process chain are only partially understood or not known at all. To meet the demand and to meet observable trends, production processes must be made ready in the shortest possible time and need to be operated efficiently. In this context, the use of virtual production system respective digital twins enables the processes of battery cell production to be optimized along the entire process chain to achieve an increase in efficiency. This paper presented the design and structure of a virtual production system for battery cell manufacturing. Using the process step of cell assembly as an example, the corresponding data model was explained and insights into the process simulation were provided. In the simulation model of the assembly process, the machine behavior and the resulting material behavior were mapped. This provides information on the energy demands of the process, the throughput and the possibility of deriving corresponding qualitative intermediate product properties. A virtual exploration of different setting parameters on the machine and material parameters can be carried out. Recommendations for action can thus be derived for each new production scenario.

In a next step, corresponding simulation/prediction models and data structures of the other processes have to be provided. Connections of the individual processes are made by the input and output parameters of the models. These models are integrated into the presented platform with the respective interfaces. Virtual production scenarios can be executed via the control and optimization system and parameter relationships can be identified. Employing of the systematic training of an AI model, the optimal parameters at the individual processes are to be determined for each production specification, according to selected criteria, and transferred to the real plants in a final step. This provides optimization potential, especially when commissioning processes in response to changing production specifications. Furthermore, the outlook is to link the simulation models directly with the control system of the respective machine. This makes it possible to use virtual sensors to record process variables that cannot be measured but are highly relevant for determining process quality. Model-based control of the individual processes is therefore possible. This also allows optimization approaches to be identified during operation. This would require the provision of models that are capable of being operated in near-real time. The transfer to reduced order models is the focus here.

Acknowledgements. The authors would like to express their appreciation to all industry partners, research partners and the German Federal Ministry of Education and Research for supporting the project ViPro (grant number: 03XP0324C) and InZePro-Cluster. This work contributes to the research performed at KIT-BATEC (KIT Battery Technology Center) and at CELEST (Center for Electrochemical Energy Storage Ulm Karlsruhe).

References

1. Michaelis, S., et al.: Roadmap Batterie-Produktionsmittel 2030: Update 2020. VDMA Verlag GmbH, Frankfurt am Main (2020)
2. Thomitzek, M.: Simulating process-product interdependencies in battery production systems. In: 51st Conference on Manufacturing Systems (2018)
3. Hettesheimer, T., Thielmann, A., Neef, C., Möller, K., Wolter, M., Lorentz, M.G.V., Wenger, M., Prill, T., Zausch, J., Kitzler, P., Montnacher, J., Miller, M., Hagen, M., Fanz, P., Tübke, J.: Entwicklungsperspektiven für Zellformate von Lithium-Ionen-Batterien in der Elektromobilität. Fraunhofer-Allianz Batterien, Pfinztal (2017)
4. Kurzweil, P., Dietlmeier, O.: Elektrochemische Speicher: Superkondensatoren, Batterien, Elektrolyse-Wasserstoff, Rechtliche Rahmenbedingungen, 2nd edn. Springer Fachmedien Wiesbaden GmbH, Wiesbaden (2018)
5. Thielmann, A., et al.: Hochenergie-Batterien 2030 + und Perspektiven zukünftiger Batterietechnologien". Fraunhofer-Institut für System- und Innovationsforschung IS **I**, 2017 (2017)
6. Asef, P., Milan, M., Lapthorn, A., Padmanaban, S.: Future trends and aging analysis of battery energy storage systems for electric vehicles. Sustainability **13**, 13779 (2021). https://doi.org/10.3390/su132413779
7. Homan, B., Kortenaar, M., Hurink, J., Smit, G.: A realistic model for battery state of charge prediction in energy management simulation tools. Energy **171**, 205–217 (2019). https://doi.org/10.1016/j.energy.2018.12.134
8. Merkle, L., Pöthig, M., Schmid, F.: Estimate e-Golf battery state using diagnostic data and a digital twin. Batteries **7**(1), 15 (2021). https://doi.org/10.3390/batteries7010015
9. Li, L., Ren, Y., O'Regan, K., Koleti, U., Kendrick, E., Widanage, D., Marco, J.: Lithium-ion battery cathode and anode potential observer based on reduced-order electrochemical single particle model. J. Energy Storage **44** (2021). https://doi.org/10.1016/j.est.2021.103324
10. Li, W., Rentemeister, M., Badeda, J., Jöst, D., Schulte, D., Sauer, D.: Digital twin for battery systems: Cloud battery management system with online state-of-charge and state-of-health estimation. J. Energy Storage **30**, 101557 (2020). https://doi.org/10.1016/j.est.2020.101557
11. Talkhestani, B., Jazdi, N., Schlögl, W., Weyrich, M.: A concept in synchronization of virtual production system with real factory based on anchor-point method. Procedia CIRP **67**, 13–17 (2018). https://doi.org/10.1016/j.procir.2017.12.168
12. Schönemann, M., Bockholt, H., Thiede, S., Kwade, A., Herrmann, C.: Multiscale simulation approach for production systems. Int. J. Adv. Manuf. Technol. **102**(5–8), 1373–1390 (2019). https://doi.org/10.1007/s00170-018-3054-y

13. Ngandjong, A.C., Lombardo, T., Primo, E.N., Chouchane, M., Shodiev, A., Arcelus, O., Franco, A.A.: Investigating electrode calendering and its impact on electrochemical performance by means of a new discrete element method model: Towards a digital twin of Li-Ion battery manufacturing. J. Power Sources **485**(1), 229320 (2021). doi: https://doi.org/10.1016/j.jpowsour.2020.229320
14. Schreiner, D., Klinger, A., Reinhart, G.: Modeling of the Calendering Process for Lithium-Ion Batteries with DEM Simulation. Procedia CIRP **93**, 149–155 (2020). doi: https://doi.org/10.1016/j.procir.2020.05.158.
15. Mayer D., Fleischer, J.: Concept for modelling the influence of electrode corrugation after calendering on stacking accuracy in battery cell production. Procedia CIRP, **104**(4), 744–749 (2021). doi: https://doi.org/10.1016/j.procir.2021.11.125.
16. Park, J., Kim, K.T., Oh, D.Y., Jin, D., Kim, D., Jung, Y.S., Lee, Y.M.: Digital Twin-Driven All-Solid-State Battery: Unraveling the Physical and Electrochemical Behaviors. Adv. Energy Mater., **10**(35), 2001563 (2020). doi: https://doi.org/10.1002/aenm.202001563.
17. Wu, B., Widanage, W.D., Yang, S., Liu, X.: Battery digital twins: Perspectives on the fusion of models, data and artificial intelligence for smart battery management systems. Energy and AI, **1**, 100016 (2020). doi: https://doi.org/10.1016/j.egyai.2020.100016.
18. Singh, S., Weeber, M., Birke, K.P.: Implementation of battery digital twin: approach, functionalities and benefits. Batteries, **7**(4), 78 (2021). doi: https://doi.org/10.3390/batteries7040078.
19. Husseini, K., Schmidgruber, N., Weinmann, H.W., Maibaum, K., Ruhland, J., Fleischer, J.: Development of a digital twin for improved Ramp-Up processes in the context of Li-Ion-Battery-Cell-Stack-Formation. Procedia CIRP, **106**(35), 27–32 (2022). doi: https://doi.org/10.1016/j.procir.2022.02.150.
20. Schel, D., Henkel, C., Stock, D., Meyer, O., Rauhöft, G., Einberger, P., Stöhr, M., Daxer, M.A., Seidelmann, J.: Manufacturing service bus: an implementation. Procedia CIRP, **67**, 179–184 (2018). doi: https://doi.org/10.1016/j.procir.2017.12.196.
21. Holtewert, P., Wutzke, R., Seidelmann, J., Bauernhansl, T.: Virtual Fort Knox Federative, secure and cloud-based platform for manufacturing. Procedia CIRP **7**, 527–532 (2013). https://doi.org/10.1016/j.procir.2013.06.027
22. Weinmann, H.W., Eichelkraut, M., Wolke da Silva, L., Fleischer, J.: Batteriezellenfertigung vom Coil zum Stack: Integriert, automatisiert und dadurch hoch flexible. C2 Coating & Converting **4**, 21–24 (2020)
23. Weinmann, H.W., Töpper, H., Fleischer, J.: Coil2Stack: Ein innovatives Verfahren zur formatflexiblen Batteriezellherstellung. ZWF Zeitschrift für wirtschaftlichen Fabrikbetrieb **115**(4), 241–243 (2020)
24. Modelica Association: Functional Mock-up Interface (FMI) (Online). https://fmi-standard.org/. Accessed 22 Feb 2022
25. Prabhakar R., Pagilla and Karl N. Reid: Governing equations for Web Tension and Web Velocity in the presence of nonideal rollers. J. Dyn. Syst. Meas. Control, **135**, (2013). doi: https://doi.org/10.1115/1.4007974.
26. Weinmann, H.W., Lang, F., Hofmann, J., Fleischer, J.: Bahnzugkraftregelung in der batteriezellfertigung. wt Werkstattstechnik online, **108**(7/8), 519–524 (2018).

Author Index

K. Dröder and T. Vietor (Eds.): *Future Automotive Production Conference 2022,*
Zukunftstechnologien für den multifunktionalen Leichtbau, pp. 265–266, 2023.
https://doi.org/10.1007/978-3-658-39928-3